If I always told you the truth,
I wouldn't need you to trust me.

LESSWRONG

Refining the art of human rationality

Trust – Introduction

The first book is about trust, the belief in something in the absence of understanding.

The Royal Society was founded in 1660, with the motto Nullius in Verba, "On No One's Word". The promise of science was that it would be a method for discovering truth without needing to trust anyone: because experiments can and would be replicated, incorrect theories and incorrect information would be filtered out. By exploring systems that could be understood in full, scientists eventually made breakthroughs in our understanding of every part of the natural world.

You might hope that this sort of understanding can be expanded further — that you could come to understand every part of the world, look inside to see how it works, and personally arrive at true beliefs about it. This hope would be wrong. To understand everything in the world would take an amount of time many orders of magnitude more than historical human lifespans.

In place of being able to understand any system, we have its lowly sibling, trust. Can I believe what this paper is telling me, when I don't have time to replicate its results or even read every page? Can I trust my emotions when I don't know where they come from? Some code on StackExchange I haven't read? A trained machine learning system off-distribution? These are questions where we can't feasibly get a repeatable, replicable setup.

The essays in this book explore the judgments we make about how to decide which sources to trust, when building our models of the world and acting upon it.

Artwork

Both the covers and all the interior artworks for each essay were generated using a machine learning system that implemented the VQGAN+CLIP algorithm for image generation. For each image the system took as input the Mississippi River covers from the previous LessWrong annual collection "A Map that Reflects the Territory", as well as a text prompt that described the artistic style and central themes. For each essay, the title was used in the prompt.

To achieve this, Asimov Collective developed their own scalable image generation service called "Multivac Art", which allows for the parallel generation of hundreds of images in an enormous variety of artistic styles and content prompts, by spinning up many hundreds of instances in parallel in the Amazon Web Services cloud.

The prompt for the cover to this book was:

The Engines of Cognition by Alex Hillkurtz | System of Gears | Aquarelle | Greek Architecture | Blue on White Color Palette | Trending on Artstation

Trust

Scott Alexander

Rule Thinkers In, Not Out

Imagine a black box which, when you pressed a button, would generate a scientific hypothesis. 50% of its hypotheses are false; 50% are true hypotheses as game-changing and elegant as relativity. Even despite the error rate, it's easy to see this box would quickly surpass space capsules, da Vinci paintings, and printer ink cartridges to become the most valuable object in the world. Scientific progress on demand, and all you have to do is test some stuff to see if it's true? I don't want to devalue experimentalists. They do great work. But it's appropriate that Einstein is more famous than Eddington. If you took away Eddington, someone else would have tested relativity; the bottleneck is in Einsteins. Einstein-in-a-box at the cost of requiring two Eddingtons per insight is a heck of a deal.

What if the box had only a 10% success rate? A 1% success rate? My guess is: still most valuable object in the world. Even a 0.1%

success rate seems pretty good, considering (what if we ask the box for cancer cures, then test them all on lab rats and volunteers?) You have to go pretty low before the box stops being great.

I thought about this after reading this list of geniuses with terrible ideas.[1] Linus Pauling thought Vitamin C cured everything. Isaac Newton spent half his time working on weird Bible codes. Nikola Tesla pursued mad energy beams that couldn't work. Lynn Margulis revolutionized cell biology by discovering mitochondrial endosymbiosis, but was also a 9-11 truther and doubted HIV caused AIDS. Et cetera. Obviously this should happen. Genius often involves coming up with an outrageous idea contrary to conventional wisdom and pursuing it obsessively despite naysayers. But nobody can have a 100% success rate. People who do this successfully sometimes should also fail at it sometimes, just because they're the kind of person who attempts it at all. Not everyone fails. Einstein seems to have batted a perfect 1000 (unless you count his support for socialism). But failure shouldn't surprise us.

Yet aren't some of these examples unforgiveably bad? Like, seriously Isaac – Bible codes? Well, granted, Newton's chemical experiments may have exposed him to a little more mercury than can be entirely healthy. But remember: gravity was considered creepy occult pseudoscience by its early enemies. It subjected the earth and the heavens to the same law, which shocked 17th century sensibilities the same way trying to link consciousness and matter would today. It postulated that objects could act on each other through invisible forces at a distance, which was equally outside the contemporaneous Overton Window. Newton's exceptional genius, his exceptional ability to think outside all relevant boxes, and his exceptionally egregious mistakes are all the same phenomenon (plus or minus a little mercury).

Or think of it a different way. Newton stared at problems that had vexed generations before him, and noticed a subtle pattern everyone else had missed. He must have amazing hypersensitive pattern-matching going on. But people with such hypersensitivity should be most likely to see patterns where they don't exist. Hence, Bible codes.

These geniuses are like our black boxes: generators of brilliant ideas, plus a certain failure rate. The failures can be easily discarded: physicists were able to take up Newton's gravity without wasting time on his Bible codes. So we're right to treat geniuses as valuable in the same way we would treat those boxes as valuable.

This goes not just for geniuses, but for anybody in the idea industry. Coming up with a genuinely original idea is a rare skill, much harder than judging ideas is. Somebody who comes up with one good original idea (plus ninety-nine really stupid cringeworthy takes) is a better use of your reading time than somebody who reliably never gets anything too wrong, but never says anything you find new or surprising. Alyssa Vance calls this positive selection[2] – a single good call rules you in – as opposed to negative selection, where a single bad call rules you out. You should practice positive selection for geniuses and other intellectuals.

I think about this every time I hear someone say something like "I lost all respect for Steven Pinker after he said all that stupid stuff about AI". Your problem was thinking of "respect" as a relevant predicate to apply to Steven Pinker in the first place. Is he your father? Your youth pastor? No? Then why are you worrying about whether or not to "respect" him? Steven Pinker is a black box who occasionally spits out ideas, opinions, and arguments for you to evaluate. If some of them are arguments you wouldn't have come up with on your own, then he's doing you a service. If 50% of them are false, then the best-case scenario is that they're moronically, obviously false, so that you can reject them quickly and get on with your life.

I don't want to take this too far. If someone has 99 stupid ideas and then 1 seemingly good one, obviously this should increase your probability that the seemingly good one is actually flawed in a way you haven't noticed. If someone has 99 stupid ideas, obviously this should make you less willing to waste time reading their other ideas to see if they are really good. If you want to learn the basics of a field you know nothing about, obviously read a textbook. If you don't trust your ability to figure out when people are wrong, obviously read someone with a track record of always representing the conventional wisdom correctly. And if you're a social engineer trying to recommend what other people who are less intelligent than you should read, obviously steer them away from anyone who's wrong too often. I just worry too many people wear their social engineer hat so often that they forget how to take it off, forget that "intellectual exploration" is a different job than "promote the right opinions about things" and requires different strategies.

But consider the debate over "outrage culture". Most of this focuses on moral outrage. Some smart person says something we consider evil, and so we stop listening to her or giving her a platform. There are arguments for and against this – at the very least it disincentivizes evil-seeming statements.

But I think there's a similar phenomenon that gets less attention and is even less defensible – a sort of intellectual outrage culture.

"How can you possibly read that guy when he's said [stupid thing]?" I don't want to get into defending every weird belief or conspiracy theory that's ever been [stupid thing]. I just want to say it probably wasn't as stupid as Bible codes. And yet, Newton. Some of the people who have most inspired me have been inexcusably wrong on basic issues. But you only need one world-changing revelation to be worth reading.

 Originally published 26th February 2019

(1) *Examples Of Otherwise Good Scientists Who Were Consumed By Their Pet Theories Despite Weak/Contradictory Evidence?* *Reddit; r/slatestarcodex. 2018.*

(2) *Alyssa Vance. Negative and Positive Selection. 5th Jul 2012.*

John S Wentworth

Gears vs Behavior

Thank you to Sisi Cheng[1] (of the Working as Intended[2] comic) for the excellent drawings.

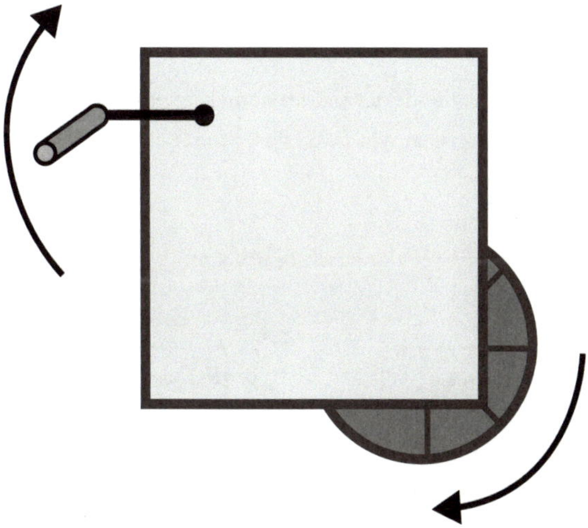

Suppose we have a gearbox. On one side is a crank, on the other side is a wheel which spins when the crank is turned. We want to predict the rotation of the wheel given the rotation of the crank, so we run a Kaggle competition.[3]

We collect hundreds of thousands of data points on crank rotation and wheel rotation. 70% are used as training data, the other 30% set aside as test data and kept under lock and key in an old nuclear bunker. Hundreds of teams submit algorithms to predict wheel rotation from crank rotation. Several top teams combine their models into one gradient-boosted deep random neural support vector forest. The model achieves stunning precision and accuracy in predicting wheel rotation.

On the other hand, in a very literal sense, the model contains no gears.[4] Is that a problem? If so, when and why would it be a problem?

What is Missing?

When we say the model "contains no gears",[4] what does that mean, in a less literal and more generalizable sense?

Simplest answer: the deep random neural support vector forest model does not tell us what we expect to see if we open up the physical gearbox.

For instance, consider these two gearboxes:

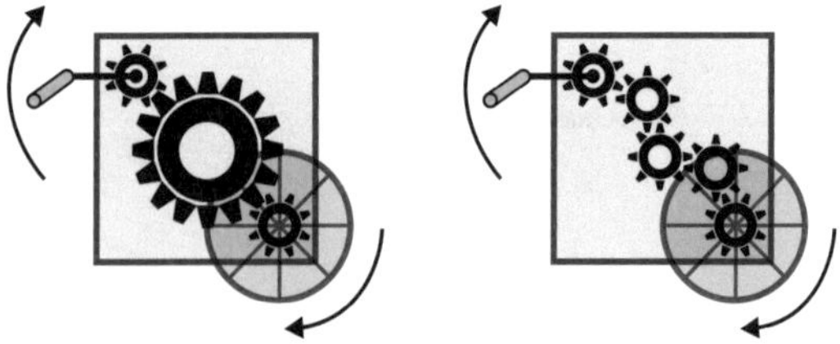

Both produce the same input-output behavior. Our model above, which treats the gearbox as a literal black box, does not tell us anything at all which would distinguish between these two cases. It only talks about input-output behavior, without making any predictions about what's inside the gearbox (other than that the gearbox must be consistent with the input/output behavior).

That's the key feature of gears-level models: they make falsifiable predictions about the internals of a system, separate from the externally-visible behavior. If a model could correctly predict all of a system's externally-visible behavior, but still be falsified by looking inside the box, then that's a gears-level model. Conversely, we cannot fully learn gears-level models by looking only at externally-visible input-output behavior - external behavior cannot, for example, distinguish between the 3- and 5-gear models above. A model which can be fully learned from system behavior, without any side information, is not a full gears-level model.

Why would this be useful, if what we really care about is the externally-visible behavior? Several reasons:

- First and foremost, if we are able to actually look inside the box, then that provides a huge amount of information about the behavior. If we can see the physical gears, then we can immediately make highly confident predictions about system behavior.
- More generally, any information about the internals of the system provides a "side channel" for testing gears-level models. If data about externally-visible behavior is limited, then the ability to leverage data about system internals can be valuable.
- It may be that all of our input data is only from within a certain range - i.e. we never tried cranking the box faster than a human could crank. If someone comes along and attaches a motor to the crank, then that's going to generate input way outside the range of what our input/output model has ever seen - but if we know what the gears look like, then that won't be a problem. In other words, knowing what the system internals look like lets us deal with distribution shifts.

- Finally, if someone changes something about the system, then a model trained only on input/output data will fail completely. For instance, maybe there's a switch on top of the gearbox which disconnects the gears, and nobody has ever thrown it before. If we know what the inside of the box looks like, then that's not a problem - we can look at what the switch does.

All that said, if we have abundant data and aren't worried about distribution shifts or system changes, non-gears models can still give us great predictive power. Solomonoff induction[5] is the idealized theoretical example: it gives asymptotically optimal predictions based on input-output behavior, without any visibility into the system internals.

Application: Macroeconomic Models

One particularly well-known example of these ideas in action is the Lucas Critique,[6] a famous 1976 paper by Bob Lucas[7] critiquing the use of simple statistical models for evaluation of economic policy decisions. Lucas' paper gives several broad examples, but arguably the most remembered example is policy decisions based on the Phillips curve.[8]

The Phillips curve is an empirical relationship between unemployment and inflation. Phillips examined almost a century of economic data, and showed a consistent negative correlation: when inflation was high, unemployment was low, and vice-versa. In other words, prices and wages rise faster at the peak of the business cycle (when unemployment is low) than at the trough (when unemployment is high).

The obvious mistake one might make, based on the Phillips curve, is to think that perpetual low unemployment can be achieved simply by creating perpetual inflation (e.g. by printing money). Lucas opens his critique by eviscerating this very idea:

> The inference that permanent inflation will therefore induce a permanent economic high is no doubt [...] ancient, yet it is only recently that this notion has undergone the mysterious transformation from obvious fallacy to cornerstone of the theory of economic policy.

Bear in mind that this was written in the mid-1970's - the era of "stagflation", when both inflation and unemployment were high for several years. Stagflation was an empirical violation of the Phillips curve - the historical behavior of the system broke down when central banks changed their policies to pursue more inflation, and people changed their behavior to account for faster expected inflation in the future.

In short: a statistical model with no gears in it completely fell apart when one part of the system (the central bank) changed its behavior.

On the other hand, before stagflation was under way, multiple theorists (notably Edmund Phelps[9] and Milton Friedman,[10] via very different approaches) published simple gears-level models of the Phillips curve which predicted that it would break down if currencies were *predictably* devalued - i.e. if people *expected* central banks to print more money. The key "gears" in these models were individual agents - the macroeconomic behavior (unemployment-inflation relationship) was explained in terms of the expectations and decisions of all the individual people.

This led to a paradigm shift in macroeconomics, beginning the era of "microfoundations": macroeconomic models derivable from microeconomic models of the expectations and behavior of individual agents - in other words, gears-level models of the economy.

Gears from Behavior?

In general, we cannot *fully* learn gears-level models by looking only at externally-visible input-output behavior. Our hypothetical 3- or 5-gear boxes are a case in point.

However, some kinds of models can at least deduce *something* about gears-level structure by looking at externally-visible behavior.

For example: given a gearbox with a crank and wheel, it's entirely possible that the rotation of the wheel has hysteresis, a.k.a. memory - it depends not only on the crank's rotation now, but also the crank's rotation earlier. This would be the case if, for instance, the box contains a flywheel. If we look at the data and see that the wheel's rotation has no dependence on the crank's rotation at earlier times (after accounting for the crank's current rotation), then we can conclude that the box probably does not contain any flywheels or other hysteretic components (or if it does, they're small or decoupled from the wheel).

More generally, these sorts of conditional independence relationships fall under the umbrella of probabilistic causal models.[11] By testing different causal models on externally-visible data, we can back out information about the internal cause-and-effect structure of the system. If we see that only the crank's current rotation matters to the wheel, then that rules out internal components with memory.

Causal models are the largest class of statistical models I know of which yield information about internal gears. However, they're not the only way to build gears-level models from behavior. If we have strong prior information, we can often use behavioral data to directly compare gears-level hypotheses.

Application: Wolf's Dice

Around the mid-19th century, Swiss astronomer Rudolf Wolf rolled a pair of dice 20,000 times, recording each outcome. The main result was that the dice were definitely not perfectly fair - there were small but statistically significant biases.

Now, we could easily look at Wolf's data and use it to estimate the frequency with which each face of each die is rolled. But that's not a gears-level model; it doesn't say anything about the physical die.

In order to back out gears-level information from the data, we need to leverage our prior knowledge about dice and die-making. Jaynes did exactly this in a 1979 paper;[12] the key pieces of prior information are:

- We know dice are roughly cube-shaped, and any difference in face frequencies should stem from asymmetry of the physical die. We know 3 is opposite 4, 2 is opposite 5, and 1 is opposite 6.
- We know dice have little pips showing the numbers on each face; different faces have different numbers of pips, which we'd expect to introduce a slight asymmetry.
- Imagining how the dice might have been manufactured, Jaynes guesses that the final cut would have been more difficult to make perfectly even than the earlier cuts - leaving one axis slightly shorter/longer than the other two.

Based on those asymmetries, we'd guess:

- One of the three face pairs (3, 4), (2, 5), (1, 6) has significantly different frequency from the others, corresponding to the last axis cut.
- The faces with fewer pips (3, 2, and especially 1) have slightly lower frequency than those with more pips (4, 5, and especially 6), since more pips means slightly less mass near that face.
- Other than that, the frequencies should be pretty even.

This is basically a guess-and-check process: we guess what asymmetry might be present based on our prior knowledge, consider how that would change the behavior, then we use the data to check the model.

Jaynes tests out these models, and finds that (1) the white die's 3-4 axis is slightly shorter than the other two, and (2) the pips indeed shift the center of mass slightly away from the center of the die. These two asymmetries together explain all of the bias seen in the data, so the die should be quite symmetric otherwise. I analyze the same problem in this post[13] (using slightly different methods from Jaynes) and reproduce the same result.

Because this is a gears-level model, we could in principle check the result using a "side channel": if we could track down the dice Wolf used, then we could take out our calipers and measure the lengths of the 3-4, 2-5, and 1-6 axes. Our prediction is that the 2-5 and 1-6 axes would be close, but the 3-4 axis would be significantly shorter. Note that we still don't have a *full* gears-level model - we don't predict *how much* shorter the 3-4 axis is. We don't have a way to back out all the dimensions of the die. But we certainly expect the difference between the 3-4 length and the 2-5 length to be much larger than the difference between the 2-5 length and the 1-6 length. Our model yields *some* information about gears-level structure.

Takeaway

Statistics, machine learning, and adjacent fields tend to have a myopic focus on predicting future data.

Gears-level models cannot be fully learned by looking at externally-visible behavior data. That makes it hard to prove theorems about convergence of statistical methods, or write tests for machine learning algorithms, when the goal is to learn about a system's internal gears. So, to a large extent, these fields have ignored gears-level learning and focused on predicting future data. Gears have snuck in only to the extent that they're useful for predicting externally-visible behavior.

But sooner or later, any field dominated by a gears-less worldview will have its Lucas Critique.

It is possible to leverage probability to test gears-level models, and to back out at least some information about a system's internal structure. It's not easy. We need to restrict ourselves to certain classes of models (i.e. causal models) and/or leverage lots of prior knowledge (e.g. about dice). It looks less like black-box statistical/ML models, and more like science: think about what the physical system might look like, figure out how the data would differ between different possible physical systems, and then go test it. The main goal is not to predict future data, but to compare models.

That's the kind of approach we need to build models which won't fall apart every time central banks change their policies.

Originally published 19th September 2019

(1) *Sisi Cheng (LessWrong users profile).*

(2) *workingasintendedcomic (Instagram profile).*

(3) *kaggle.com. Competitions.*

(4) *Valentine. Gears in understanding. 11th May 2017.*

(5) *Ming Li, Paul M.B. Vitányi. An Introduction to Kolmogorov Complexity and Its Applications (Texts in Computer Science). 21st Nov 2008.*

(6) *Robert E. Lucas, Jr. Econometric Policy Evaluation: A Critique.*

(7) *Wikipedia. Robert Lucas Jr.*

(8) *Wikipedia. Phillips curve.*

(9) *Wikipedia. Edmund Phelps.*

(10) *Wikipedia. Milton Friedman.*

(11) *EliezerYudkowsky. Causal Diagrams and Causal Models. 12th Oct 2012*

(12) *E.T Jaynes. Concentrations Of Distributions At Entropy Maxima.*

(13) *John S Wentworth. Wolf's Dice II: What Asymmetry? 17th Jul 2019.*

Scott Alexander

Book Review: The Secret Of Our Success

I.

"Culture is the secret of humanity's success" sounds like the most vapid possible thesis. The Secret Of Our Success,[1] by anthropologist Joseph Heinrich, manages to be an amazing book anyway.

Heinrich wants to debunk (or at least clarify) a popular view where humans succeeded because of our raw intelligence. In this view, we are smart enough to invent neat tools that help us survive and adapt to unfamiliar environments.

Against such theories: we cannot actually do this. Heinrich walks the reader through many stories about European explorers marooned in unfamiliar environments. These explorers usually starved to death. They starved to death in the middle of endless plenty. Some of them were in Arctic lands that the Inuit considered among their richest hunting grounds. Others were in jungles, surrounded by edible plants and animals. One particularly unfortunate group was in Alabama, and would have perished entirely if they hadn't been captured and enslaved by local Indians first.

These explorers had many advantages over our hominid ancestors. For one thing, their exploration parties were made up entirely of strong young men in their prime, with no need to support women, children, or the elderly. They were often selected for their education and intelligence. Many of them were from Victorian Britain, one of the most successful civilizations in history, full of geniuses like Darwin and Galton. Most of them had some past experience with wilderness craft and survival. But despite their big brains, when faced with the task our big brains supposedly evolved for – figuring out how to do hunting and gathering in a wilderness environment – they failed pathetically.

Nor is it surprising that they failed. Hunting and gathering is actually really hard. Here's Heinrich's description of how the Inuit hunt seals:

> You first have to find their breathing holes in the ice. It's important that the area around the hole be snow-covered—otherwise the seals will hear you and vanish. You then open the hole, smell it to verify it's still in use (what do seals smell like?), and then assess the shape of the hole using a special curved piece of caribou antler. The hole is then covered with snow, save for a small gap at the top that is capped with a down indicator. If the seal enters the hole, the indicator moves, and you must

blindly plunge your harpoon into the hole using all your weight. Your harpoon should be about 1.5 meters (5ft) long, with a detachable tip that is tethered with a heavy braid of sinew line. You can get the antler from the previously noted caribou, which you brought down with your driftwood bow.

The rear spike of the harpoon is made of extra-hard polar bear bone (yes, you also need to know how to kill polar bears; best to catch them napping in their dens). Once you've plunged your harpoon's head into the seal, you're then in a wrestling match as you reel him in, onto the ice, where you can finish him off with the aforementioned bear-bone spike.

Now you have a seal, but you have to cook it. However, there are no trees at this latitude for wood, and driftwood is too sparse and valuable to use routinely for fires. To have a reliable fire, you'll need to carve a lamp from soapstone (you know what soapstone looks like, right?), render some oil for the lamp from blubber, and make a wick out of a particular species of moss. You will also need water. The pack ice is frozen salt water, so using it for drinking will just make you dehydrate faster. However, old sea ice has lost most of its salt, so it can be melted to make potable water. Of course, you need to be able to locate and identify old sea ice by color and texture. To melt it, make sure you have enough oil for your soapstone lamp.

No surprise that stranded explorers couldn't figure all this out. It's more surprising that the Inuit did. And although the Arctic is an unusually hostile place for humans, Heinrich makes it clear that hunting-gathering techniques of this level of complexity are standard everywhere. Here's how the Indians of Tierra del Fuego make arrows:

Among the Fuegians, making an arrow requires a 14-step procedure that involves using seven different tools to work six different materials. Here are some of the steps:

- The process begins by selecting the wood for the shaft, which preferably comes from chaura, a bushy, evergreen shrub. Though strong and light, this wood is a non-intuitive choice since the gnarled branches require extensive straightening (why not start with straighter branches?).
- The wood is heated, straightened with the craftsman's teeth, and eventually finished with a scraper. Then, using a pre-heated and grooved stone, the shaft is pressed into the grooves and rubbed back and forth, pressing it down with a piece of fox skin. The fox skin become impregnated with the dust, which prepares it for the polishing stage (Does it have to be fox skin?).
- Bits of pitch, gathered from the beach, are chewed and mixed with ash (What if you don't include the ash?).
- The mixture is then applied to both ends of a heated shaft, which must then be coated with white clay (what about red clay? Do you have to heat it?). This prepares the ends for the fletching and arrowhead.
- Two feathers are used for the fletching, preferably from upland geese (why not chicken feathers?).
- Right-handed bowman must use feathers from the left wing of the bird, and vice versa for lefties (Does this really matter?).
- The feathers are lashed to the shaft using sinews from the back of the guanaco, after they are smoothed and thinned with water and saliva (why not sinews from the fox that I had to kill for the aforementioned skin?).
- Next is the arrowhead, which must be crafted and then attached to the shaft, and of course there is also the bow, quiver and archery skills. But, I'll leave it there, since I think you get the idea.

How do hunter-gatherers know how to do all this? We usually summarize it as "culture". How did it form? Not through some smart Inuit or Fuegian person reasoning it out; if that had been it, smart European explorers should have been able to reason it out too.

The obvious answer is "cultural evolution",[2] but Heinrich isn't much better than anyone else at taking the mystery out of this phrase. Trial and error must have been involved, and less successful groups/people imitating the techniques of more successful ones. But is that really a satisfying explanation?

I found the chapter on language a helpful reminder that we already basically accept something like this is true. How did language get invented? I'm especially interested in this question because of my brief interactions with conlanging communities – people who try to construct their own languages as a hobby or as part of a fantasy universe, like Tolkien did with Elvish. Most people are terrible at this; their languages are either unusable, or exact clones of English. Only people who (like Tolkien) already have years of formal training in linguistics can do a remotely passable job. And you're telling me the original languages were invented by cavemen? Surely there was no committee of Proto-Indo-European nomads that voted on whether to have an inflecting or agglutinating tongue? Surely nobody ran out of their cave shouting "Eureka!" after having discovered the interjection? We just kind of accept that after cavemen worked really hard to communicate with each other, eventually language – still one of the most complicated and impressive productions of the human race – just sort of happened.

Taking the generation of culture as secondary to this kind of mysterious process, Heinrich turns to its transmission. If cultural generation happens at a certain rate, then the fidelity of transmission determines whether a given society advances, stagnates, or declines.

For Heinrich, humans started becoming more than just another species of monkey when we started transmitting culture with high fidelity. Some anthropoligsts talk about the Machiavellian Intelligence Hypothesis[3] – the theory that humans evolved big brains in order to succeed at social manuevering and climbing dominance hierarchies. Heinrich counters with his own Cultural Intelligence Hypothesis – humans evolved big brains in order to be able to maintain things like Inuit seal hunting techniques. Everything that separates us from the apes is part of an evolutionary package designed to help us maintain this kind of culture, exploit this kind of culture, or adjust to the new abilities that this kind of culture gave us.

II.

Secret gives many examples of many culture-related adaptations, and not all are in the brain.

One of the most important differences between man and ape is our puny digestive tracts:

> Our mouths are the size of the squirrel monkey's, a species that weighs less than three pounds. Chimpanzees can open their mouths twice as wide as we can and hold substantial amounts of food compressed between their lips and large teeth. We also have puny jaw muscles that reach up only to just below our ears. Other primates' jaw muscles stretch to the top of their heads, where they sometimes even latch onto a central bony ridge. Our stomachs are small, having only a third of the surface area that we'd expect for a primate of our size, and our colons are too short, being only 60% of their expected mass.

Compared to other animals, we have such atrophied digestive tracts that we shouldn't be able to live. What saves us? All of our food processing techniques, especially cooking, but also chopping, rinsing, boiling, and soaking. We've done much of the work of digestion before food even enters our mouths. Our culture teaches us how to do this, both in broad terms like "hold things over fire to cook them" and in specific terms like "this plant needs to be soaked in water for 24 hours to leach out the toxins". Each culture has its own cooking knowledge related to the local plants and animals; a frequent cause of death among European explorers was cooking things in ways that didn't unlock any of the nutrients, and so starving while apparently well-fed.

All of this is cultural. Heinrich is kind of cruel in his insistence on this. He recommends readers go outside and try to start a fire. He even gives some helpful hints – flint is involved, rubbing two sticks together works for some people, etc. He predicts – and stories I've heard from unfortunate campers confirm – that you will not be able to do this, despite an IQ far beyond that of most of our hominid ancestors. In fact, some groups (most notably the aboriginal Tasmanians) seem to have lost the ability to make fire, and never rediscovered it. Fire-making was discovered a small number of times, maybe once, and has been culturally transmitted since then.

And food processing techniques are even more complicated. Nixtamalization of corn, necessary to prevent vitamin deficiencies, involves soaking the corn in a solution containing ground-up burnt seashells. The ancient Mexicans discovered this and lived off corn just fine for millennia. When the conquistadors took over, they ignored it and ate corn straight. For four hundred years, Europeans and Americans ate unnixtamalized corn. By official statistics, three million Americans came down with corn-related vitamin deficiencies during this time, and up to a hundred thousand died. It wasn't until 1937 that Western scientists discovered which vitamins were involved and developed an industrial version of nixtamalization that made corn safe. Early 1900s Americans were very smart and had lots of advantages over ancient Mexicans. But the ancient Mexicans' culture got this one right in a way it took Westerners centuries to match.

Humans are persistence hunters: they cannot run as fast as gazelles, but they can keep running for longer than gazelles (or almost anything else). Why did we evolve into that niche? The secret is our ability to carry water. Every hunter-gatherer culture has invented its own water-carrying techniques, usually some kind of waterskin. This allowed humans to switch to perspiration-based cooling systems, which allowed them to run as long as they want.

And humans are consumate tool users. In some cases, we evolved in order to use tools better; our hands outclass those of any other ape in terms of finesse. In other cases, we devolved systems that were no longer necessary once tools took over. We are vastly weaker than any other ape. Heinrich describes a circus act of the 1940s where the ringmaster would challenge strong men in the audience to wrestle a juvenile chimpanzee. The chimpanzee was tied up, dressed in a mask that prevented it from biting, and wearing soft gloves that prevented it from scratching. No human ever lasted more than five seconds. Our common ancestor with other apes grew weaker and weaker as we became more and more reliant on artificial weapons to give us an advantage.

III.

But most of our differences from other apes are indeed in the brain. They're just not necessarily where you would expect.

Tomasello et al. tested human toddlers vs. apes on a series of traditional IQ type questions. The match-up was surprisingly fair; in areas like memory, logic, and spatial reasoning, the three species did about the same. But in ability to learn from another person, humans wiped the floor with the other two ape species:

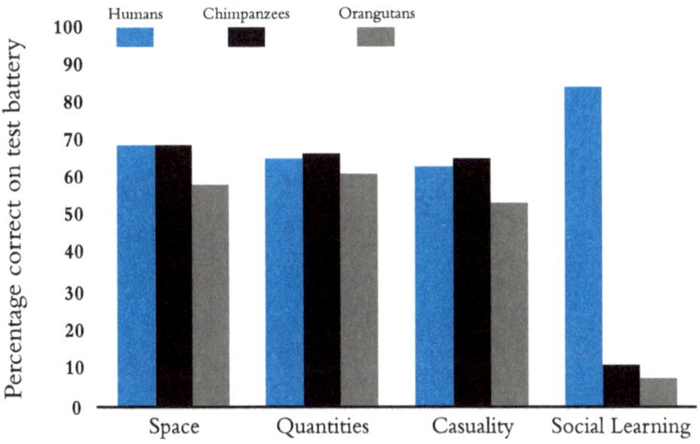

Average performance on four sets of cognitives tests with toddlers, chimpanzees and orangutans.

Remember, Heinrich thinks culture accumulates through random mutation. Humans don't have control over how culture gets generated. They have more control over how much of it gets transmitted to the next generation. If 100% gets transmitted, then as more and more mutations accumulate, the culture becomes better and better. If less than 100% gets transmitted, then at some point new culture gained and old culture lost fall into equilibrium, and your society stabilizes at some higher or lower technological level. This means that transmitting culture to the next generation is maybe the core human skill. The human brain is optimized to make this work as well as possible.

Human children are obsessed with learning things. And they don't learn things randomly. There seem to be "biases in cultural learning", ie slots in an infant's mind that they know need to be filled with knowledge, and which they preferentially seek out the knowledge necessary to fill.

One slot is for language. Human children naturally listen to speech (as early as in the womb). They naturally prune the phonemes they are able to produce and distinguish to the ones in the local language. And they naturally figure out how to speak and understand what people are saying, even though learning a language is hard even for smart adults.

Another slot is for animals. In a world where megafauna has been relegated to zoos, we still teach children their ABCs with "L is for lion" and "B is for bear", and children still read picture books about Mr. Frog and Mrs. Snake holding tea parties. Heinrich suggests that just as the young brain is hard-coded to want to learn language, so it is hard-coded to want to learn the local animal life (little boys' vehicle obsession may be a weird outgrowth of this; buses and trains are the closest thing to local megafauna that most of them will encounter).

Another slot is for plants:

> To see this system in operation, let's consider how infants respond to unfamiliar plants. Plants are loaded with prickly thorns, noxious oils, stinging nettles and dangerous toxins, all genetically evolved to prevent animals like us from messing with them. Given our species wide geographic range and diverse use of plants as foods, medicines and construction materials, we ought to be primed to both learn about plants and avoid their dangers. To explore this idea in the lab, the psychologists Annie Wertz and Karen Wynn first gave infants, who ranged in age from

eight to eighteen months, an opportunity to touch novel plants (basil and parsley) and artifacts, including both novel objects and common ones, like wooden spoons and small lamps.

The results were striking. Regardless of age, many infants flatly refused to touch the plants at all. When they did touch them, they waited substantially longer than they did with the artifacts. By contrast, even with the novel objects, infants showed none of this reluctance. This suggests that well before one year of age infants can readily distinguish plants from other things, and are primed for caution with plants. But, how do they get past this conservative predisposition?

The answer is that infants keenly watch what other people do with plants, and are only inclined to touch or eat the plants that other people have touched or eaten. In fact, once they get the 'go ahead' via cultural learning, they are suddenly interested in eating plants. To explore this, Annie and Karen exposed infants to models who both picked fruit from plants and also picked fruit-like things from an artifact of similar size and shape to the plant. The models put both the fruit and the fruit-like things in their mouths. Next, the infants were given a choice to go for the fruit (picked from the plant) or the fruit-like things picked from the object. Over 75% of the time the infants went for the fruit, not the fruit-like things, since they'd gotten the 'go ahead' via cultural learning.

As a check, the infants were also exposed to models putting the fruit or fruit-like things behind their ears(not in their mouths). In this case, the infants went for the fruit or fruit-like things in equal measure. It seems that plants are most interesting if you can eat them, but only if you have some cultural learning cues that they aren't toxic.

After Annie first told me about her work while I was visiting Yale in 2013, I went home to test it on my 6-month-old son, Josh. Josh seemed very likely to overturn Annie's hard empirical work, since he immediately grasped anything you gave him and put it rapidly in his mouth. Comfortable in his mom's arms, I first offered Josh a novel plastic cube. He delighted in grabbing it and shoving it directly into his mouth, without any hesitation. Then, I offered him a sprig of arugula. He quickly grabbed it, but then paused, looked with curious uncertainty at it, and then slowly let it fall from his hand while turning to hug his mom.

It's worth pointing out how rich the psychology is here. Not only do infants have to recognize that plants are different from objects of similar size, shape and color, but they need to create categories for types of plants, like basil and parsley, and distinguish 'eating' from just 'touching'. It does them little good to code their observation of someone eating basil as 'plants are good to eat' since that might cause them to eat poisonous plants as well as basil. But, it also does them little good to narrowly code the observation as 'that particular sprig of basil is good to eat' since that particular sprig has just been eaten by the person they are watching. This is another content bias in cultural learning.

This ties into the more general phenomenon of figuring out what's edible. Most Westerners learn insects aren't edible; some Asians learn that they are. This feels deeper than just someone telling you insects aren't edible and you believing them. When I was in Thailand, my guide offered me a giant cricket, telling me it was delicious. I believed him when he said it was safe to eat, I even believed him when he said it tasted good to him, but my conditioning won out – I didn't eat the cricket. There seems to be some process where a child's brain learns what is and isn't locally edible, then hard-codes it against future change.

(Or so they say; I've never been able to eat shrimp either).

Another slot is for gender roles. By now we've all heard the stories of progressives who try to raise their children without any exposure to gender. Their failure has sometimes been taken as evidence that gender is hard-coded. But it can't be quite that simple: some modern gender roles, like girls = pink, are far from obvious or universal. Instead, it looks like children have a hard-coded slot that gender roles go into, work hard to figure out what the local gender roles are (even if their parents are trying to confuse them), then latch onto them and don't let go.

In the Cultural Intelligence Hypothesis, humans live in obligate symbiosis with a culture. A brain without an associated culture is incomplete and not very useful. So the infant brain is adapted to seek out the important aspects of its local culture almost from birth and fill them into the appropriate slots in order to become whole.

IV.

The next part of the book discusses post-childhood learning. This plays an important role in hunter-gatherer tribes:

> While hunters reach their peak strength and speed in their twenties, individual hunting success does not peak until around age 30, because success depends more on know-how and refined skills than on physical prowess.

This part of the book made most sense in the context of examples like the Inuit seal-hunting strategy which drove home just how complicated and difficult hunting-gathering was. Think less "Boy Scouts" and more "PhD"; a primitive tribesperson's life requires mastery of various complicated technologies and skills. And the difference between "mediocre hunter" and "great hunter" can be the difference between high status (and good mating opportunities) and low status, or even between life and death. Hunter-gatherers really want to learn the essentials of their hunter-gatherer lifestyle, and learning it is really hard. Their heuristics are:

Learn from people who are good at things and/or widely-respected. If you haven't already read about the difference between dominance and prestige hierarchies, check out Kevin Simler's blog post[4] on the topic. People will fear and obey authority figures like kings and chieftains, but they give a different kind of respect ("prestige") to people who seem good at things. And since it's hard to figure out who's good at things (can a non-musician who wants to start learning music tell the difference between a merely good performer and one of the world's best?) most people use the heuristic of respecting the people who other people respect. Once you identify someone as respect-worthy, you strongly consider copying them in, well, everything:

To understand prestige as a social phenomenon, it's crucial to realize that it's often difficult to figure out what precisely makes someone successful. In modern societies, the success of a star NBA basketball player might arise from his:

(1) intensive practice in the offseason
(2) sneaker preference
(3) sleep schedule
(4) pre-game prayer
(5) special vitamins
(6) taste for carrots

Any or all of these might increase his success. A naïve learner can't tell all the causal links between an individual's practices and his success. As a consequence, learners often copy their chosen models broadly across many domains. Of course, learners may place more weight on domains that for one reason or other seem more causally relevant to the model's success. This copying often includes the model's personal habits or styles as well as their goals and motivations, since these may be linked to their success. This "if in doubt, copy it" heuristic is one of the reasons why success in one domain converts to influence across a broad range of domains.

The immense range of celebrity endorsements in modern societies shows the power of prestige. For example, NBA star Lebron James, who went directly from High School to the pros, gets paid millions to endorse State Farm Insurance. Though a stunning basketball talent, it's unclear why Mr. James is qualified to recommend insurance companies. Similarly, Michael Jordan famously wore Hanes underwear and apparently Tiger Woods drove Buicks. Beyoncé drinks Pepsi (at least in commercials). What's the connection between musical talent and sugary cola beverages?

> Finally, while new medical findings and public educational campaigns only gradually influence women's approach to preventive medicine, Angelina Jolie's single OP-ED in the New York Times, describing her decision to get a preventive double mastectomy after learning she had the 'faulty' BRCA1 gene, flooded clinics from the U.K. to New Zealand with women seeking genetic screenings for breast cancer. Thus, an unwanted evolutionary side effect, prestige turns out to be worth millions, and represents a powerful and underutilized public health tool.

Of course, this creates the risk of prestige cascades, where some irrelevant factor (Heinrich mentions being a reality show star) catapults someone to fame, everyone talks about them, and you end up with Muggeridge's definition of a celebrity: someone famous for being famous.

Some of this makes more sense if you go back to the evolutionary roots, and imagine watching the best hunter in your tribe to see what his secret is, or being nice to him in the hopes that he'll take you under his wing and teach you stuff.

(But if all this is true, shouldn't public awareness campaigns that hire celebrity spokespeople be wild successes? Don't they just as often fail, regardless of how famous a basketball player they can convince to lecture schoolchildren about how Winners Don't Do Drugs)?

Learn from people who are like you. If you are a man, it is probably a bad idea to learn fashion by observing women. If you are a servant, it is probably a bad idea to learn the rules of etiquette by observing how the king behaves. People are naturally inclined to learn from people more similar to themselves.

Heinrich ties this in to various studies showing that black

students learn best from a black teacher, female students from a female teacher, et cetera.

Learn from old people. Humans are almost unique in having menopause; most animals keep reproducing until they die in late middle-age. Why does evolution want humans to stick around without reproducing?

Because old people have already learned the local culture and can teach it to others. Heinrich asks us to throw out any personal experience we have of elders; we live in a rapidly-changing world where an old person is probably "behind the times". But for most of history, change happened glacially slowly, and old people would have spent their entire lives accumulating relevant knowledge. Imagine a world where when a Silicon Valley programmer can't figure out how to make his code run, he calls up his grandfather, who spent fifty years coding apps for Google and knows every programming language inside and out.

Sometimes important events only happen once in a generation. Heinrich tells the story of an Australian aboriginal tribe facing a massive drought. Nobody knew what to do except Paralji, the tribe's oldest man, who had lived through the last massive drought and remembered where his own elders had told him to find the last-resort waterholes.

This same dynamic seems to play out even in other species:

> In 1993, a severe drought hit Tanzania, resulting in the death of 20% of the African elephant calves in a population of about 200. This population contained 21 different families, each of which was led by a single matriarch. The 21 elephant families were divided into 3 clans, and each clan shared the same territory during the wet season (so, they knew each other).

Researchers studying these elephants have analyzed the survival of the calves and found that families led by older matriarchs suffered fewer deaths of their calves during this drought.

Moreover, two of the three elephant clans unexpectedly left the park during the drought, presumably in search of water, and both had much higher survival rates than the one clan that stayed behind. It happens that these severe droughts only hit about once every four to five decades, and the last one hit about 1960. After that, sadly, elephant poaching in the 1970's killed off many of the elephants who would have been old enough in 1993 to recall the 1960 drought. However, it turns out that exactly one member of each of the two clans who left the park, and survived more effectively, were old enough to recall life in 1960. This suggests, that like Paralji in the Australian desert, they may have remembered what to do during a severe drought, and led their groups to the last water refuges. In the clan who stayed behind, the oldest member was born in 1960, and so was too young to have recalled the last major drought.

More generally, aging elephant matriarchs have a big impact on their families, as those led by older matriarchs do better at identifying and avoiding predators (lions and humans), avoiding internal conflicts and identifying the calls of their fellow elephants. For example, in one set of field experiments, researchers played lion roars from both male and female lions, and from either a single lion or a trio of lions. For elephants, male lions are much more dangerous than females, and of course, three lions are always worse than only one lion. All the elephants generally responded with more defensive preparations when they heard three lions vs. one. However, only the older matriarchs keenly recognized the increased dangers of male lions over female lions, and responded to the increased threat with elephant defensive maneuvers.

V.

I was inspired to read Secret by this review on Scholar's Stage.[5] I hate to be unoriginal, but after reading the whole book, I agree that the three sections Tanner cites – on divination, on manioc, and on shark taboos – are by far the best and most fascinating.

On divination:

> When hunting caribou, Naskapi foragers in Labrador, Canada, had to decide where to go. Common sense might lead one to go where one had success before or to where friends or neighbors recently spotted caribou.
>
> However, this situation is like [the Matching Pennies[6] game]. The caribou are mismatchers and the hunters are matchers. That is, hunters want to match the locations of caribou while caribou want to mismatch the hunters, to avoid being shot and eaten. If a hunter shows any bias to return to previous spots, where he or others have seen caribou, then the caribou can benefit (survive better) by avoiding those locations (where they have previously seen humans). Thus, the best hunting strategy requires randomizing.
>
> Can cultural evolution compensate for our cognitive inadequacies? Traditionally, Naskapi hunters decided where to go to hunt using divination and believed that the shoulder bones of caribou could point the way to success. To start the ritual, the shoulder blade was heated over hot coals in a way that caused patterns of cracks and burnt spots to form. This patterning was then read as a kind of map, which was held in a pre-specified orientation. The cracking patterns were (probably) essentially random from the point of view of hunting locations, since the

outcomes depended on myriad details about the bone, fire, ambient temperature, and heating process. Thus, these divination rituals may have provided a crude randomizing device that helped hunters avoid their own decision-making biases.

This is not some obscure, isolated practice, and other cases of divination provide more evidence. In Indonesia, the Kantus of Kalimantan use bird augury to select locations for their agricultural plots. Geographer Michael Dove argues that two factors will cause farmers to make plot placements that are too risky. First, Kantu ecological models contain the Gambler's Fallacy, and lead them to expect floods to be less likely to occur in a specific location after a big flood in that location (which is not true). Second…Kantus pay attention to others' success and copy the choices of successful households, meaning that if one of their neighbors has a good yield in an area one year, many other people will want to plant there in the next year. To reduce the risks posed by these cognitive and decision-making biases, Kantu rely on a system of bird augury that effectively randomizes their choices for locating garden plots, which helps them avoid catastrophic crop failures. Divination results depend not only on seeing a particular bird species in a particular location, but also on what type of call the bird makes (one type of call may be favorable, and another unfavorable).

The patterning of bird augury supports the view that this is a cultural adaptation. The system seems to have evolved and spread throughout this region since the 17th century when rice cultivation was introduced. This makes sense, since it is rice cultivation that is most positively influenced by randomizing garden locations. It's possible that, with the introduction of rice, a few farmers began to use bird sightings as an indication of favorable garden sites. On-average, over a lifetime, these farmers would do better – be more successful – than farmers

> who relied on the Gambler's Fallacy or on copying others' immediate behavior. Whatever the process, within 400 years, the bird augury system spread throughout the agricultural populations of this Borneo region. Yet, it remains conspicuously missing or underdeveloped among local foraging groups and recent adopters of rice agriculture, as well as among populations in northern Borneo who rely on irrigation. So, bird augury has been systematically spreading in those regions where it's most adaptive.

Scott Aaronson has written about how easy it is to predict people trying to "be random":

> In a class I taught at Berkeley, I did an experiment where I wrote a simple little program that would let people type either "f" or "d" and would predict which key they were going to push next. It's actually very easy to write a program that will make the right prediction about 70% of the time. Most people don't really know how to type randomly. They'll have too many alternations and so on. There will be all sorts of patterns, so you just have to build some sort of probabilistic model. Even a very crude one will do well. I couldn't even beat my own program, knowing exactly how it worked. I challenged people to try this and the program was getting between 70% and 80% prediction rates. Then, we found one student that the program predicted exactly 50% of the time. We asked him what his secret was and he responded that he "just used his free will."

But being genuinely random is important in pursuing mixed game theoretic strategies. Heinrich's view is that divination solved this problem effectively.

I'm reminded of the Romans using augury to decide when and where to attack. This always struck me as crazy; generals are going to risk the lives of thousands of soldiers because they saw a weird bird earlier that morning? But war is a classic example of when a random strategy can be useful. If you're deciding whether to attack the enemy's right vs. left flank, it's important that the enemy can't predict your decision and send his best defenders there. If you're generally predictable – and Scott Aaronson says you are – then outsourcing your decision to weird birds might be the best way to go.

And then there's manioc. This is a tuber native to the Americas. It contains cyanide, and if you eat too much of it, you get cyanide poisoning. From Heinrich:

> In the Americas, where manioc was first domesticated, societies who have relied on bitter varieties for thousands of years show no evidence of chronic cyanide poisoning. In the Colombian Amazon, for example, indigenous Tukanoans use a multistep, multiday processing technique that involves scraping, grating, and finally washing the roots in order to separate the fiber, starch, and liquid. Once separated, the liquid is boiled into a beverage, but the fiber and starch must then sit for two more days, when they can then be baked and eaten. Figure 7.1 shows the percentage of cyanogenic content in the liquid, fiber, and starch remaining through each major step in this processing.
>
> Such processing techniques are crucial for living in many parts of Amazonia, where other crops are difficult to cultivate and often unproductive. However, despite their utility, one person would have a difficult time figuring out the detoxification technique. Consider the situation from the point of view of the children and adolescents who are learning the techniques. They would have rarely, if ever, seen anyone get cyanide poisoning, because the techniques work. And even if the processing was ineffective, such that cases of goiter (swollen necks) or neurological problems were common, it would still be hard to recognize the link between these chronic health issues and eating manioc.

Most people would have eaten manioc for years with no apparent effects. Low cyanogenic varieties are typically boiled, but boiling alone is insufficient to prevent the chronic conditions for bitter varieties. Boiling does, however, remove or reduce the bitter taste and prevent the acute symptoms (e.g., diarrhea, stomach troubles, and vomiting).

So, if one did the common-sense thing and just boiled the high-cyanogenic manioc, everything would seem fine. Since the multistep task of processing manioc is long, arduous, and boring, sticking with it is certainly non-intuitive. Tukanoan women spend about a quarter of their day detoxifying manioc, so this is a costly technique in the short term. Now consider what might result if a self-reliant Tukanoan mother decided to drop any seemingly unnecessary steps from the processing of her bitter manioc. She might critically examine the procedure handed down to her from earlier generations and conclude that the goal of the procedure is to remove the bitter taste. She might then experiment with alternative procedures by dropping some of the more labor-intensive or time-consuming steps. She'd find that with a shorter and much less labor-intensive process, she could remove the bitter taste. Adopting this easier protocol, she would have more time for other activities, like caring for her children. Of course, years or decades later her family would begin to develop the symptoms of chronic cyanide poisoning.

Thus, the unwillingness of this mother to take on faith the practices handed down to her from earlier generations would result in sickness and early death for members of her family. Individual learning does not pay here, and intuitions are misleading. The problem is that the steps in this procedure are causally opaque—an individual cannot readily infer their functions, interrelationships, or importance. The causal opacity of many cultural adaptations had a big impact on our psychology.

> Wait. Maybe I'm wrong about manioc processing. Perhaps it's actually rather easy to individually figure out the detoxification steps for manioc? Fortunately, history has provided a test case. At the beginning of the seventeenth century, the Portuguese transported manioc from South America to West Africa for the first time. They did not, however, transport the age-old indigenous processing protocols or the underlying commitment to using those techniques. Because it is easy to plant and provides high yields in infertile or drought-prone areas, manioc spread rapidly across Africa and became a staple food for many populations. The processing techniques, however, were not readily or consistently regenerated. Even after hundreds of years, chronic cyanide poisoning remains a serious health problem in Africa. Detailed studies of local preparation techniques show that high levels of cyanide often remain and that many individuals carry low levels of cyanide in their blood or urine, which haven't yet manifested in symptoms. In some places, there's no processing at all, or sometimes the processing actually increases the cyanogenic content. On the positive side, some African groups have in fact culturally evolved effective processing techniques, but these techniques are spreading only slowly.

Rationalists always wonder: how come people aren't more rational? How come you can prove a thousand times, using Facts and Logic, that something is stupid, and yet people will still keep doing it?

Heinrich hints at an answer: for basically all of history, using reason would get you killed.

A reasonable person would have figured out there was no way for oracle-bones to accurately predict the future. They would have abandoned divination, failed at hunting, and maybe died of starvation.

A reasonable person would have asked why everyone was wasting so much time preparing manioc. When told "Because that's how we've always done it", they would have been unsatisfied with that answer. They would have done some experiments, and found that a simpler process of boiling it worked just as well. They would have saved lots of time, maybe converted all their friends to the new and easier method. Twenty years later, they would have gotten sick and died, in a way so causally distant from their decision to change manioc processing methods that nobody would ever have been able to link the two together.

Heinrich discusses pregnancy taboos in Fiji; pregnant women are banned from eating sharks. Sure enough, these sharks contain chemicals that can cause birth defects. The women didn't really know why they weren't eating the sharks, but when anthropologists demanded a reason, they eventually decided it was because their babies would be born with shark skin rather than human skin. As explanations go, this leaves a lot to be desired. How come you can still eat other fish? Aren't you worried your kids will have scales? Doesn't the slightest familiarity with biology prove this mechanism is garbage? But if some smart independent-minded iconoclastic Fijian girl figured any of this out, she would break the taboo and her child would have birth defects.

In giving humans reason at all, evolution took a huge risk. Surely it must have wished there was some other way, some path that made us big-brained enough to understand tradition, but not big-brained enough to question it. Maybe it searched for a mind design like that and couldn't find one. So it was left with this ticking time-bomb, this ape that was constantly going to be able to convince itself of hare-brained and probably-fatal ideas. Here, too, culture came to the rescue. One of the most important parts of any culture – more important than the techniques for hunting seals, more important than the techniques for processing tubers – is techniques for making sure nobody ever questions tradition. Like the belief that anyone who doesn't conform is probably a witch who should be cast out lest they bring destruction upon everybody. Or the belief in a God who has commanded certain specific weird dietary restrictions, and will torture you forever if you disagree. Or the fairy tales where the prince asks a wizard for help, and the wizard says "You may have everything you wish forever, but you must never nod your head at a badger", and then one day the prince nods his head at a badger, and his whole empire collapses into dust, and the moral of the story is that you should always obey weird advice you don't understand.

There's a monster at the end of this book. Humans evolved to transmit culture with high fidelity. And one of the biggest threats to transmitting culture with high fidelity was Reason. Our ancestors lived in Epistemic Hell, where they had to constantly rely on causally opaque processes with justifications that couldn't possibly be true, and if they ever questioned them then they might die. Historically, Reason has been the villain of the human narrative, a corrosive force that tempts people away from adaptive behavior towards choices that "sounded good at the time".

Why are people so bad at reasoning? For the same reason they're so bad at letting poisonous spiders walk all over their face without freaking out. Both "skills" are really bad ideas, most of the people who tried them died in the process, so evolution removed those genes from the population, and successful cultures stigmatized them enough to give people an internalized fear of even trying.

VI.

This book belongs alongside Seeing Like A State[7] and the works of G.K. Chesterton[8] as attempts to justify tradition, and to argue for organically-evolved institutions over top-down planning. What unique contribution does it make to this canon?

First, a lot more specifically anthropological/ paleoanthropological rigor than the other two.

Second, a much crisper focus: Chesterton had only the fuzziest idea that he was writing about cultural evolution, and Scott was only a little clearer. I think Heinrich is the only one of the three to use the term, and once you hear it, it's obviously the right framing.

Third, a sense of how traditions contain the meta-tradition of defending themselves against Reason, and a sense for why this is necessary.

And fourth, maybe we're not at the point where we really want unique contributions yet. Maybe we're still at the point where we have to have this hammered in by more and more examples. The temptation is always to say "Ah, yes, a few simple things like taboos against eating poisonous plants may be relics of cultural evolution, but obviously by now we're at the point where we know which traditions are important vs. random looniness, and we can rationally stick to the important ones while throwing out the garbage." And then somebody points out to you that actually divination using oracle bones was one of the important traditions, and if you thought you knew better than that and tried to throw it out, your civilization would falter.

Maybe we just need to keep reading more similarly-themed books until this point really sinks in, and we get properly worried.

(1) *Joseph Henrich, The Secret of Our Success: How Culture is Driving Human Evolution, Domesticating Our Species, and Making Us Smarter. 2015.*

(2) *birguslatro, Why Cultural Evolution Is Real (And What It Is). 22nd Nov 2014.*

(3) *Lydia M Hopper, Erica van de Waal, & Christine A Caldwell, Celebrating the continued importance of "Machiavellian Intelligence" 30 years on. 2018.*

(4) *Kevin Simler, Social Status: Down the Rabbit Hole. 13th Oct 2015.*

(5) *Tanner Greer, Tradition is Smarter Than You Are. 27th Aug 2018.*

(6) *Wikipedia.org/Matching_pennies*

(7) *Scott Alexander, Book Review: Seeing Like a Slate. 16 Mar 2017.*

(8) *Wikipedia.org/Chesteron%27s_fence*

Ben Hoffman

Reason isn't magic

Here's a story some people like to tell[1] about the limits of reason. There's this plant, manioc, that grows easily in some places and has a lot of calories in it, so it was a staple for some indigenous South Americans since before the Europeans showed up. Traditional handling of the manioc involved some elaborate time-consuming steps that had no apparent purpose, so when the Portuguese introduced it to Africa, they didn't bother with those steps - just, grow it, cook it, eat it.

The problem is that manioc's got cyanide in it, so if you eat too much too often over a lifetime, you get sick, in a way that's not easily traceable to the plant. Somehow, over probably hundreds of years, the people living in manioc's original range figured out a way to leach out the poison, without understanding the underlying chemistry - so if you asked them why they did it that way, they wouldn't necessarily have a good answer.

Now a bunch of Africans growing and eating manioc as a staple regularly get cyanide poisoning.

This is offered as a cautionary tale against innovating through reason, since there's a lot of information embedded in your culture (via hundreds of years of selection), even if people can't explain why. The problem with this argument is that it's a nonsense comparison.

First of all, it's not clear things got worse on net, just that a tradeoff was made. How many person-days per year were freed up by less labor-intensive manioc handling? Has anyone bothered to count the hours lost to laborious traditional manioc-processing, to compare them with the burden of consuming too much cyanide? How many of us, knowing that convenience foods probably lower our lifespans relative to slow foods, still eat them because they're... more convenient?

How many people *didn't starve* because manioc was available and would grow where and when other things wouldn't?

If this is the best we can do for how poorly reason can perform, reason seems pretty great.

Second, we're not actually comparing reason to tradition - we're comparing *changing things* to *not changing things*. Change, as we know, is bad.[2] Sometimes we change things anyway - when we think it's worth the price, or the risk. Sometimes, we're wrong.

Third, the actually existing Portuguese and Africans involved in this experiment weren't committed rationalists - they were just people trying to get by. It probably doesn't take more than a day's reasoning to figure out which steps in growing manioc are really necessary to get the calories palatably. Are we imagining that someone making a concerted effort to improve their life through reason would just stop there?

This is being compared with *many generations* of trial and error. Is that the standard we want to use? Reasoning isn't worth it unless a day of untrained thinking can outperform hundreds of years of accumulated tradition?

It gets worse. This isn't a randomly selected example - it's specifically selected as a case where reason would have a hard time noticing when and how it's making things worse. In this particular case, reason introduced an important problem. But life is full of risks, sometimes in ways that are worse for traditional cultures. Do we really want to say that reasoning isn't the better bet unless it outperforms literally every time, without ever making things locally worse? Even theoretically perfect Bayesian rationality will sometimes recommend changes that have an expected benefit, but turn out to be harmful. Not even tradition meets this standard! Only logical certainties do - provided, that is, we haven't made an error in one of our proofs.

We also have to count all the deaths and other problems averted by reasoning about a problem. Reasoning introduces risks - but also, risks come up even when we're not reasoning about them, just from people doing things that affect their environments. There's absolutely no reason to think that the sort of gradual iteration that accretes into tradition never enters a bad positive feedback loop. Even if you think modernity is an exceptional case of that kind of bad feedback loop, we had to have gotten there via the accretion of premodern tradition and iteration! The only way out is through. But why did we have this exaggerated idea of what reason could do, in the first place?

(1) *Scott Alexander, Book Review: The Secret of Our Success. 4 Jun 2019.*

(2) *Zvi Mowshowitz, Change Is Bad. 20 Jul 2017.*

"Other people are wrong" vs "I am right"

I've recently been spending some time thinking about the rationality mistakes I've made in the past. Here's an interesting one: I have historically been too hasty to go from "other people seem very wrong about this topic" to "I am right about this topic".

Throughout my life, I've often thought that other people had beliefs that were really repugnant and stupid. Now that I am older and wiser, I still think that these ideas were repugnant and stupid. Overall, I was probably not quite dismissive enough of things like the opinions of apparent domain experts and the opinions of people who seemed smart but whose arguments I couldn't really follow. I also overrated conventional wisdom about factual claims about how the world worked, though I underrated conventional wisdom about how to behave.

Examples of ideas where I thought the conventional wisdom was really dumb:

- I thought that animal farming was a massive moral catastrophe, and I thought it was a sign of terrible moral failure that almost everyone around me didn't care about this and wasn't interested when I brought it up.

- I thought that AI safety was a big deal, and I thought the arguments against it were all pretty stupid. (Nowadays the conventional wisdom has a much higher opinion of AI safety; I'm talking about 2010-2014).

- I thought that people had terrible taste in economic policy, and that they mostly voted for good-sounding stuff that stops sounding good if you think about it properly for even a minute.

- I was horrified by people proudly buying products that said "Made in Australia" on them; I didn't understand how that wasn't obviously racist, and I thought that we should make it much easier to allow anyone who wants to to come live in Australia. (This one has become much less controversial since Trump inadvertently convinced liberals that they should be in favor of immigration liberalization).

- I thought that a lot of people's arguments about why it's good to call the police on bike thieves were dumb. See e.g. many of the arguments people made in response to a post of mine about this.[1] (that, in fairness, was a really dumb post IMO).

I think I was right about other people being wrong. However, I think that my actual opinions on these topics were *also* pretty confused and wrong, much more than I thought at the time. Here's how I updated my opinions for all the things above:

- I have updated against the simple view of hedonic utilitarianism under which it's plausible that simple control systems can suffer.

A few years ago, I was seriously worried that the future would contain much more factory farming and therefore end up net negative; I now think that I overrated this fear, because (among other arguments) almost no one actually endorses torturing animals, we just do it out of expediency, and in the limit of better technology our weak preferences will override our expediency.

- My understanding of AI safety was "eventually someone will build a recursively self improving singleton sovereign AGI, and we need to figure out how to build it such that it can have an off switch and it implements some good value function instead of something bad." I think this picture was massively oversimplified. On the strategic side, I didn't think about the possibilities of slower takeoffs or powerful technologies without recursive self improvement; on the technical safety side, I didn't understand that it's hard to even build a paperclip maximizer,[2] and a lot of our effort might go into figuring out how to do that.

- Other people have terrible taste in economic policy, but I think that I was at the time overconfident in various libertarian-ish ideas that I'm now less enthusiastic about. Also, I no longer think it's a slam dunk that society is better off from becoming wealthier, because of considerations related to the far future, animals, and whether more money makes us happier.

- I think that immigration liberalization is more dangerous than I used to think, because rich societies seem to generate massive positive externalities for the rest of the world and it seems possible that a sudden influx of less educated people with (in my opinion) worse political opinions might be killing the goose that lays the golden eggs.

- On bike thieves, I think that even though utilitarianism is good and stuff, it's extremely costly to have thievery be tolerated, because then you have to do all these negative-sum things like buying bike locks. Also it seems like we're generally better off if people help with enforcement of laws.

• • •

In all of these cases, my arguments against others were much higher quality than my actual beliefs. Much more concerningly, I think I was significantly better at spotting the holes in other people's arguments than spotting holes in my own.

There's also a general factor here of me being overconfident in the details of ideas that have some ring of truth to them. Like, the importance of AGI safety seemed really obvious to me, and I think that my sense of obviousness has historically been pretty good at spotting arguments that later stand up to intense scrutiny. But I was massively overconfident in my particular story for how AGI would go down. I should have been more disjunctive: I should have said "It sure seems like something like this ought to happen, and it seems like step three could happen in any of these four possible ways, and I don't know which of them will be true, and maybe it will actually be another one, but I feel pretty convinced that there's some way it will happen".

Here are some other ideas which I continue to endorse which had that ring of truth to them, but the details of which I've been similarly overconfident about. (Some of these are pretty obscure.)

- The simulation hypothesis[3]
- UDASSA[4]
- The malignancy of the universal prior[5]
- The mathematical universe hypothesis[6]
- Humans have weird complex biases related to categories like race and gender, and we should be careful about this in our

thinking. (Nowadays this idea is super widespread and so it feels weird to put it in the same list as all these crazy other ideas. But when I first encountered it seriously in my first year of college, it felt like an interesting and new idea, in the same category as many of the cognitive biases I heard about on LessWrong).

And here are ideas which had this ring of truth to them that I no longer endorse:

- We should fill the universe with hedonium.
- The future might be net negative, because humans so far have caused great suffering with their technological progress and there's no reason to imagine that this will change. Futurists are biased against this argument because they personally don't want to die and have a strong selfish desire for human civilization to persist.
- Because of Landauer's limit,[7] civilizations have an incentive to aestivate. (This one is wrong because it involves a misunderstanding of thermodynamics).[8]

• • •

My bias towards thinking my own beliefs are more reasonable than they are would be disastrous if it prevented me from changing my mind in response to good new arguments. Luckily, I don't think that I am particularly biased in that direction, for two reasons. Firstly, when I'm talking to someone who thinks I'm wrong, for whatever reason I usually take them pretty seriously, and I have a small crisis of faith that prompts me to go off and reexamine my beliefs a bunch. Secondly, I think that most of the times when someone has presented an argument which later changed my mind, my initial reaction has been confusion rather than dismissiveness.

As an example of the first: Once upon a time I told someone who I respected that they shouldn't eat animal products, because of the vast suffering caused by animal farming. He looked over scornfully and told me that it was pretty rich for me to say that, given that I use Apple products—hadn't I heard about the abusive Apple factory conditions and how they have nets to prevent people killing themselves by jumping off the tops of the factories? I felt terrified that I'd been committing some grave moral sin, and then went off to my room to research the topic for an hour or two. I eventually became convinced that the net effect of buying Apple products on human welfare is probably very slightly positive but small enough to not worry about, and also it didn't seem to me that there's a strong deontological argument against doing it.

(I went back and told the guy about the result of me looking into it. He said he didn't feel interested in the topic anymore and didn't want to talk about it. I said "wow, man, I feel pretty annoyed by that; you gave me a moral criticism and I took it really seriously; I think it's bad form to not spend at least a couple minutes hearing about what I found". Someone else who was in the room, who was very enthusiastic about social justice, came over and berated me for trying to violate someone else's preferences about not talking about something. I learned something that day about how useful it is to take moral criticism seriously when it's from people who don't seem very directed by their morals).

Other examples: When I first ran across charismatic people who were in favor of deontological values and social justice-y beliefs, I took those ideas really seriously and mulled them over a lot. A few weeks ago, someone gave me some unexpectedly harsh criticism about my personal manner and several aspects of how I approach my work; I updated initially quite far in the direction of their criticism, only to update 70% of the way back towards my initial views after I spent ten more hours thinking and talking to people about it.

Examples of the second: When I met people whose view of AI safety didn't match my own naive view, I felt confused and took them seriously (including when they were expressing a bunch of skepticism of MIRI). When my friend Howie told me he thought the criminal justice system was really racist, I was surprised and quickly updated my opinion to "I am confused about this", rather than dismissing him.

I can't think of cases where I initially thought an argument was really stupid but then it ended up convincing either me or a majority of people who I think of as my epistemic peers or superiors (e.g. people who I think have generally good judgement at EA orgs).

I can, however, think of cases where I felt initially that an argument was dumb, but lots of my epistemic peers think that the argument is at least sort of reasonable. I am concerned by this and I'm trying to combat it. For example, the following arguments are in my current list of things that I am worried I'm undervaluing because they initially seem implausible to me, and are on my to-do list to eventually look into more carefully:

- Drexler's Comprehensive AI Systems
- AI safety via ambitious value learning
- Arguments that powerful AI won't lead to a singleton

Please let me know if you have examples along these lines where I seemed dumber than I'm presenting here.

• • •

Here's another perspective on why my approach might be a problem: I think that people are often pretty bad at expressing why they believe things, and in particular they don't usually say "I don't know why I believe this, but I believe it anyway". So if

I dismiss arguments that suck, I might be dismissing useful knowledge that other people have gained through experience.

I think I've made mistakes along these lines in the past. For example, I used to have a much lower opinion of professionalism than I now do. Similarly, there are a couple of serious personal mistakes I've made where I looked around for the best arguments against doing something weird I wanted to do, and all of those arguments sucked, and then I decided to do the weird thing, and then it was a bad idea.

Katja Grace calls this mistake[9] "breaking Chesterton's fence in the presence of bull".

This would suggest the heuristic "Take received wisdom on topics into account, even if you ask people where the received wisdom comes from and they tell you a source that seems extremely unreliable".

I think this heuristic is alright, but shouldn't be an overriding consideration. The ideas that evolve through the experience of social groups are valuable because they're somewhat selected for truth and importance. But the selection process for these ideas is extremely simple and dumb.

I'd expect that in most cases where something is bad, there is a legible argument for why we shouldn't do it (where I'm including arguments from empirical evidence as legible arguments). I'd prefer to just learn all of the few things that society implicitly knows, rather than giving up every time it disagrees with me.

Maybe this is me being arrogant again, but I feel like the mistake I made with the bike-stealing thing wasn't me refusing to bow to social authority, it was me not trying hard enough to think carefully about the economics of the situation. My inside view is that if I now try to think about economics, I don't need to incorporate that much outside-view-style discounting of my own arguments.

I have the big advantage of being around people who are really good at articulating the actual reasons why things are bad. Possibly the number one strength of the rationalist community is creating and disseminating good explicit models of things that are widely implicitly understood (e.g. variants of Goodhart's law,[10] Moloch,[11] Chesterton's fence,[12] the unilateralist's curse,[13] "toxoplasma of rage").[14] If I was in any other community, I'm worried that I'd make posts like the one about the bike, and no one would be able to articulate why I was wrong in a way that was convincing. So I don't necessarily endorse other people taking the strategy I take.

I am not aware of that many cases where I believed something really stupid because all the common arguments against it seemed really dumb to me. If I knew of more cases like this, I'd be more worried about this.

• • •

Claire Zabel says, in response to all this:

> I'd say you're too quick to buy a whole new story if it has the ring of truth, and too quick to ask others (and probably yourself) to either refute on the spot, or accept, a complex and important new story about something about the world, and leave too little room to say "this seems sketchy but I can't articulate how" or "I want to think about it for a while" or "I'd like to hear the critics' counterarguments" or "even though none of the above has yielded fruit, I'm still not confident about this thing".

This seems plausible. I spend a bunch of time trying to explain why I'm worried about AI risk to people who don't know much about the topic. This requires covering quite a lot of ground; perhaps I should try harder to explicitly say "by the way, I know I'm telling you a lot of crazy stuff; you should take as long as it takes to evaluate all of this on your own; my goal here is just to explain what I believe; you should use me as a data point about one place that human beliefs sometimes go after thinking about the subject".

• • •

I feel like my intuitive sense of whether someone else's argument is roughly legit is pretty good, and I plan to continue feeling pretty confident when I intuitively feel like someone else is being dumb. But I am trying to not make the jump from "I think that this argument is roughly right" to "I think that all of the steps in a more fleshed out version of that argument are roughly right". Please let me know if you think I'm making that particular mistake.

 Originally published 22nd February 2019

(1) *Buck Shlegeris. 31 July 2016.*

(2) *LessWrong, Tags. Paperclip maximizer.*

(3) *Wikipedia. Simulation hypothesis.*

(4) *http://fennetic.net/ . UDASSA.*

(5) *Paul Christiano. What does the universal prior actually look like? 30 Nov 2016*

(6) *Wikipedia. Mathematical universe hypothesis.*

(7) *Wikipedia. Landauer's principle.*

(8) *Comment on 'The aestivation hypothesis for resolving Fermi's paradox'. Charles H. Bennett, Robin Hanson, C. Jess Riedel. 18 Feb 2019.*

(9) *Mistakes #4: breaking Chesterton's fence in the presence of bull. Katja Grace. 6 Sept 2015.*

(10) *LessWrong, Tags. Goodhart's Law.*

(11) *LessWrong, Tags. Moloch.*

(12) *Wikitionary. Chesterton's fence*

(13) *Nick Bostrom, Thomas Douglas, Anders Sandberg. The Unilateralist's Curse and the Case for a Principle of Conformity. 26 Jan 2016.*

(14) *Scott Alexander. THE TOXOPLASMA OF RAGE. 17 Dec 2014.*

Duncan Sabien

In My Culture

One of my colleagues has a neat little trick. Sometimes, in the middle of a conversation—*especially* if things are heated or tense or confusing—they'll pause, and they'll say something like:

"I'm sorry, but—look. In *my* culture…"

…and then they'll go on to explain something that one might easily mistake for a rule, or a request, or the enforcement of a social norm, but which is *actually* (as far as I can tell) just a statement about the version of the world they carry around inside their head. Because none of us *quite* have the same culture, after all. We all differ in different ways from the Basic Package—even those of us who've lived in the same towns, gone to the same schools, worked in the same industries, played the same sports, read the same books, watched the same shows—we've all got our own unique little takes, built up out of the odd quirks of our parents, tiny traumas and formative experiences, countless accumulated musings about how Things Could Be So Much Better If Everyone Would Just _________!

And so Your Culture, though it might match mine at a thousand different points, will also be noticeably different at a thousand others. You and I would found different churches, write different constitutions, build different schools and startups—

—and we would recognize different things as trespasses or offenses, and react to those trespasses and offenses in different ways.

> *"They must talk to each other directly, Ender, mind to mind. What one thinks, another can also think; what one remembers, another can also remember. Why would they ever develop language? Why would they ever learn to read and write? How would they know what reading and writing were if they saw them? Or signals? Or numbers? Or anything that we use to communicate? This isn't just a matter of translating from one language to another. They don't have a language at all. We used every means we could think of to communicate with them, but they don't even have the machinery to know we're signaling. And maybe they've been trying to think to us, and they can't understand why we don't respond".*
>
> *"So the whole war is because we can't talk to each other".*
>
> *"If the other fellow can't tell you his story, you can never be sure he isn't trying to kill you".*

There's a particular school of therapy called Internal Family Systems, which was founded on the basic insight that a lot of the stuff we'd come up with for mediating conflicts between family members could also be used for handling disagreements within oneself. Identify conflicting impulses, goals, and restraints, loosely anthropomorphize them as child-figures or parent-figures or spouse-figures or sibling-figures, and treat them with the same sort of respect you'd offer real quarreling humans, and voilà—suddenly a bunch of interpersonal tools and skills can be used to make personal therapeutic progress.

In a similar fashion, my colleague has been doing something like adapting the tools of *international diplomacy* for use in difficult conversations between family and friends and coworkers. Not by demanding that everyone follow their rules, and not by asserting any one culture's superiority over another, but simply by *acknowledging the fact* that their culture meaningfully differs from everyone else's. By taking that fact *seriously*, the same way that diplomats do, and putting forth conscious effort to work sensibly around it.

I don't claim to know what's going on in my colleague's head, but from the outside, the process I see looks something like this:

- Someone will make a move that's generally known to be okay in the broader context culture, but which would be recognized as mildly offensive or blasphemous if everyone were on the exact same cultural page as my colleague (like offering a handshake to a business partner from Thailand).
- They'll pause, and do their best to make that aspect of their culture transparent and accessible and comprehensible—"in my culture, if someone is in the middle of a thought the way I just was, and somebody else interrupts in the fashion that just happened, this is generally interpreted to mean X, Y, and Z about what was going on in both our heads and what our relative positions are and so forth, and it has such-and-such ripples through the social fabric of the conversation, and as a

side note I get the sense that you expect me to be able to just pick back up in the middle of the sentence but actually this isn't a reasonable expectation because in my culture thoughts don't quite work like that." The emphasis is always on clarifying the sort of Rube Goldberg causal structure of the culture—if you push button A, you'll get effect B, and if you want effect C, you should push button D, and here's why those specific if-then relationships make sense in context.

- If necessary, they'll ask for clarification about the *other* person's culture. If, say, it wasn't entirely clear how the other person might have thought their actions were good and correct, or if there's any lingering resentment about the trespass that might be cleared up by a richer understanding of where the other person was coming from.
- Occasionally, there'll be a negotiation about which norms should be adopted in the larger context—whether side A's norms, or side B's norms, or a specific compromise between the two, or a set of triggers for distinguishing when to do A versus when to do B. More often, though, it's simply left up to everyone to do what makes sense, given *both* the knowledge that things-will-sometimes-land-this-way *and* the knowledge that this doesn't compel anyone to change their behavior in any particular way.

(It happens to be a part of this particular conversational culture that *sharing something about yourself* is distinct from *making a request of others*; i.e. you can say "X results in me being sad" without implying "if you continue to do X, you're bad/will be interpreted as intentionally trying to harm me").

• • •

I find this beautiful. Beautiful enough that it is now one of the foremost elements of my culture—that in *my* culture, this is a move people can make, and which I myself will make unilaterally. Not only does it allow me to reduce the number of unnecessary spats and misunderstandings to the theoretical minimum—

(i.e. fewer and fewer trespasses happen "for no reason"; I and my conversational partners asymptotically approach a state in which all the disagreements are *real* disagreements, and not just translation confusions over what was intended by a given hand gesture or conversational move)

—but it also gives me a glimpse into each person's vision of *utopia*. The tiny flaws that each person sees in the larger culture around them, the tiny little fixes and wishes that they've built up in themselves over the years—when I'm embedded within a subculture that does this kind of cooperative conversational anthropology on the regular, it feels like I'm constantly expanding my sense of what-better-worlds-are-possible.

And there's something particularly attractive (to me) about the frame of "this is *already* true in my culture". It's sort of like the NVC_1 shift of expressing emotions and impacts rather than claims about other people's intentions—even if you're the only person with your particular culture, you are an example of it working, and so it's less about arguing over the shared normspace and more about just laying out options and alternatives. "This is how it works for the people of Duncanland" is simultaneously more true and less confrontational than "but we all know we should X" or "obviously things would be better if Y" or "I can't *believe* you would Z"!

(I think this applies all over the place. It's often easy for requests of a cultural nature to either slide toward, or be interpreted as, something like "change your behavior to match my cultural expectations, or be seen as a bad person". And naturally, even the *perception* that that's what's happening is enough to make some people dig in their heels—from that perspective, someone's trying to move the *overall* culture to a new setpoint, and also preemptively allocating the costs of that shift. That's the sort of thing that will reliably produce resistance in *some* fraction of humans, no matter the nature or magnitude of the request.

Whereas a statement like, "Oh, in my culture we do X, and we find this low-cost and mundane" highlights an *opportunity* for greater cooperativeness, but falls short of being a demand. The frame creates a boundary right from the start—this is *my* culture, not *everyone's* culture, and therefore conformity is … contingent on relationship, or something? Like, you'll certainly need to up your game in this area if you want to do a lot of business with me, but to the extent that our interactions are confined to the broader context culture, and to the extent that the norms I'm personally comfortable with are not actually universal in that culture, it's more a matter of politeness/friendliness/extra credit. There's a strong implication that your nonconformity, as long as it's not clearly antagonistic, will be taken in good faith, just as someone from Greece will likely forgive you if you happen to forget that a thumbs-up is a rude gesture in their home culture. And surprise surprise, that gentler version is often *more* successful at getting people to willingly and cheerfully put forth effort than a version that is perceived as being tinged with demand or threat or judgment).

And sure, it's true that none of us actually have perfect introspection, and that maybe we think of things as being part of our personal culture that aren't in practice, or we mislabel and misdescribe them, and maybe there are things that kind-of-work for us as individuals that would actually be disastrous if they were installed for everyone, and so on and so on. Perhaps simply declaring "in my culture, X is true" is overconfident or misleading.But it seems much less overconfident than declaring "X is true for everyone", or "X *should* be true for everyone", and so it feels to me like a step in the right direction, even if it glosses over some important uncertainty/humility.

• • •

I want to zoom in for a moment and give some examples of the sorts of things that might fill in the blank after the words "in my culture".

- What interactions between different classes of people are or are not okay (e.g. whether a high-school teacher can or can't give their students permission to refer to them by their first name).
- What certain actions are widely interpreted to mean (e.g. "everybody knows" that if you plead the fifth, it's because you're guilty).
- What obligations you have when discussing controversial topics[2] (e.g. whether you have a responsibility to include historical context, or a responsibility to remain dispassionate and logical).
- What sorts of idiosyncrasies are defections (e.g. whether it's okay to go to a party and be the only one who remains sober).
- The "rules of engagement" for various levels of disagreement or antagonism (e.g. whether it's justified to engage in otherwise-disallowed behavior if it's in response to someone else doing it first).
- The order and ranking of swear words and epithets; which curses are mild and which are unspeakable and unforgivable.
- What things are sacred and what things may be mocked.
- Whether or not birthdays and anniversaries are special.
- When and how to decide between focusing on people's feelings and interpretations in a way that doesn't particularly care about the truth of the matter (which is often useful for healing and processing) versus focusing on cause and effect and objectively verifiable truth (which is often useful for a lot of other things).
- What it's okay for an employer to ask of its employees, and whether or not society should put limits on what adults can freely agree to (e.g. questions about mandated minimum wage).
- Which side of an argument or disagreement is the default, and

which bears the burden of proof, and how that decision is made, and whether it's made implicitly or explicitly.

- Whether or not cargo shorts are a valid thing to wear, and whether or not the question "are cargo shorts a valid thing to wear"? is a valid one to pose.

• • •

The above list was generated mostly by me trying to recall examples of times I've found myself in disagreement with other white, college-educated, non-religious, American-raised, upper-middle-class males, to show that even people who look extremely similar to me on paper can still have meaningfully different cultures. And of course, that list could easily have been hundreds or thousands of entries long.

One important thing to note is that even the smallest and most trivial entries (like the cargo shorts thing) are more than enough to send destabilizing waves rippling through the social fabric—to turn a good evening into a bad one, for instance, or to spark long-lasting antipathy between people who were otherwise getting along just fine. The TV show Seinfeld got endless mileage out of just this sort of cultural mismatch, and it was resonant and successful in part because *that sort of thing happens all the time.* It's sort of tragic, and it would be nice if we could do better, but as things currently stand it's a part of our social reality.

This is part of why the diplomatic approach gives me cautious hope. It seems like a hammer in a world full of nails—the sort of thing that might *actually* make a difference, both in the moment and in the aggregate.

In that spirit, I'd like to take the remainder of this essay to outline a dozen or so of the more important and idiosyncratic customs that are native to *my* culture. It may be that none of them seem interesting to you, or worth the effort, but it may be that some of them are small gifts you'll feel like giving me, on the level of learning to say "thank you" in someone's native tongue, and it may be that some of them are worth taking home and tossing into your own cultural stew.

One last thought, though, before diving in to my own norms and traditions: there's a difference between

noticing that there are many different cultures, and wanting to find a context culture that allows them to communicate and cooperate to the extent that there is mutual desire to do so

and

treating all cultures as fundamentally equally good or valid.

I don't want to take a firm stance on the latter question in this particular essay. I certainly have opinions, but those opinions have enough doubt and fog and uncertainty that they're not worth promoting. I simply want it noticed that you can advocate for the former (for practical reasons) *without* advocating for the latter (which seems like more of a moral or ethical question). Even when the US and the USSR were at each other's throats over a conflict of cultures, they still were able to recognize the instrumental value of having a red phone, and that seems to me like it was the right move.

• • •

In my culture, everybody simply takes for granted the existence of a larger *context* culture that exists to foster communication, cooperation, conversation, compromise, and so on and so forth. It's understood that this context culture emerged over the course of long, hard millennia of trial and error, with lots of death and suffering and war and outrage along the way, and that, while it's far from perfect, it's absolutely critical to the safety and success that those of us currently living in the developed, modern world rely upon and enjoy.

(For instance, the founding fathers of America were certainly aware of the flaws in the concept of religious freedom, but they rightly recognized that those flaws were *vastly less terrible* than the history of war and persecution and oppression that came from attempts to constrain how people worship).

It's understood that attempts to *change* this context culture are fraught with risk, and that sometimes even shining too bright of a light on its faults and failings can be a mistake. This is because, in any given room of 100 people, at least *some* will misunderstand a claim of "some part of this is bad" to mean "this whole thing is bad". And of those, some will go on to virtuously attempt to tear down the whole thing, not realizing that they're digging away at the foundation that lets us have any sort of civilization at all.

(To continue the religious freedom example, it's often not clear to a young idealist why *even saying the sentence* "hey, is religious freedom actually the right way to go"? out loud in a public forum might be treated as a risky or dangerous act, especially given *other* cornerstones of our context culture like the enshrined freedom of personal expression. But in fact the sort of unofficial social pressure that is often brought to bear to discourage such expressions is an important part of the context culture's immune system. Turns out that such sentences are often a prelude to atrocity, and even when they *aren't*, it turns out that it's reasonable for people to experience a kind of background elevated stress in response to hearing them, and even when it's not reasonable, turns out some people will *anyway*, with corresponding downstream effects, and that wishing this weren't the case or talking about how it *shouldn't* be the case doesn't actually fix it, and that much of the time we'd rather keep on quietly and imperfectly cooperating than engage with or tolerate a literally endless series of erosive complaints and suggestions).

Thus, it's known and acknowledged (in my culture) that all of the expressions below are *things which may not be possible* in the broader context culture. They are things which may be negotiated away in the service of building a world that accommodates more people. They are things which may occasionally be sacrificed, for the sake of collaboration across cultural lines. They are things which it is easy to find oneself arguing for with language that implies that *disagreement* is tantamount to *defection*—that one is either with you, and on the side of goodness and rightness, or on the *other* side—and in my culture we take the stance that letting oneself get away with that kind of slippery narrativemancy is a meaningful transgression against the larger context culture that lets us all live side by side in relative peace.

And so, *to the extent that we want to continue cooperating with people of other cultures* (i.e. to the extent that there is a diplomatic summit going on that you want to be a part of and are willing to make tradeoffs to stay a part of), people in my culture acknowledge that it is important to be careful, and to be charitable, and to prepare for one's care and charity to be insufficient, and to be wary of poking holes in the social fabric, and to be ready to repair those holes that will inevitably accidentally be poked, and to—when writing an essay such as this one—take pains to draw a bright and solid line between "this is how it is in my home culture" and "this is a bid that the overall context culture shift in my direction," and to not immediately dismiss a claim that we did not draw that line brightly and solidly enough.

Another way to put this is that, in my culture, if you want to engage in an act of civil disobedience in order to change the culture around you, it's important that you *at least stick* around and let them arrest you and put you in jail. Otherwise (so say my cultural norms) you won't be interpreted as trying to cooperatively influence that context culture, but rather as trying to outright undermine it and replace it with your own.

(Which is sometimes justified, to be clear. Sometimes you are trying to break the context culture, because the context culture is rounding up and murdering Jews and you symbolically going to jail won't actually accomplish anything. The point is not that you should always cooperate with the context culture, but that you can't have your cake and eat it too—you can *either* cooperate with it or break from it, and you should *pick*).

(It's worth noting that these points are so obvious in my culture that the first draft of this essay excluded them entirely, and my friends had to remind me to make them explicit. I just *took it for granted* that the rest of the post would be interpreted in that light, which is not at all obvious to people in cultures even only just a little bit different from mine).

• • •

In my culture, with the exception of certain subsets of neuroatypicals or disabled persons, it is assumed that the vast majority of people over the age of (say) eight or nine are *straightforwardly acting in accordance with their will.*

This is not to say that people don't make mistakes. It's not ignoring the fact that people often have their options constrained by circumstance. It's not meant to legitimize or trivialize things like oppression and coercion and abuse.

But *in general, for most people, most of the time,* the assumption is that you are doing things mostly on purpose, and that the sum of your inner values is *revealed* through your visible actions—that you are pretty much always choosing your most-preferred option from the options available. In my culture, there isn't really any such thing as *akrasia*, or a persistent inability to act in line with your values. It's understood that people often have *aspirational* values—things which they genuinely wish they could express in their actions even though they currently kind of don't—but in general if you're not acting in line with those, it's assumed that it's because you have some other (possibly hidden) wants that are getting in the way. Thus, in my culture, it's considered reasonable and standard for people to model you and react to you as if you generally want and endorse the things you generally do. Your *revealed preferences* are taken to be a reflection of your net values in all but a few special cases. Hence, apologies are treated as falling somewhere between predictions and promises—if you are *genuinely* sorry, then in my culture you will demonstrate it through future action that meaningfully differs from past action. And if your future actions continue to resemble the past, people in my culture will conclude both that you *weren't actually sorry* and also perhaps that your word can't really be trusted in general.

(Another way to say this is that, in my culture, you get to decide what's true *inside your head*, but other people are justified and supported in making predictions and taking actions based on what you *do in the world*. If you have a mood disorder that makes you judgmental and suspicious and unpredictably mean twenty percent of the time, you're welcome to say "that's not really me", but other people are *not* mistreating you or morally bankrupt if they decide not to take the risk and end up avoiding you altogether—if, in essence, they *behave toward you* as if that really *is* part of your character).

• • •

In my culture, it's understood that emotions can be right or wrong—or, to use less charged language, they can be *appropriate responses* to a situation, or they can be *inappropriate responses*. In the latter case, it is absolutely still a sacred right to *have* the emotion (no one in my culture can tell you that you should feel a different way than you feel), but what you *do* with that emotion is constrained by social norms.

This means that if Alex does something, and Bailey becomes angry, and furthermore if both Alex and Bailey want to avoid such situations in the future, the fix could be on *either* side. It's neither obviously on Alex to stop doing the thing, nor obviously on Bailey to leash their anger. Instead, the question at hand becomes something like "what actually happened? What are the relevant contextual details? How reasonable is it for Bailey to feel a sense of trespass at Alex's actions? Was a commonly-acknowledged boundary violated? What about a rarely-acknowledged-but-previously-communicated-and-agreed-upon one? Sometimes people jump to conclusions, or make wrong guesses at motives, or get angry at a made-up caricature of the other person—is that happening in this case, or is Bailey's anger a straightforward response to Alex's observable actions"?

This may sound trivial or obvious, but it's important to note that what this means, in practice, is that sometimes *I will feel wronged, and I will act in what feels like defense of myself or of important principles, and afterward I will have to apologize (and possibly accept punishment), because it will be deemed that my sense-of-being-wronged and my subsequent actions were not, in fact, an appropriate or justified reaction to the situation.* In my culture, when we find ourselves on the wrong side of judgment in this way, we often suck it up and make public peace even if we still hold resentment or uncertainty on the inside. Because otherwise—if your peers can't ask for surface-level amends unless and until you've made a deep, internal update—you could simply get away with anything by sort of ... weaponizedly never-noticing-you-were-wrong?

To put it another way, in my culture, sometimes one's personal satisfaction must come second to a sort of public dance which maintains the norms and standards of the group as a whole, and those norms and standards are seen as sensible and necessary and worthy of occasional personal sacrifice.

• • •

In my culture, there are *limits* to the judgments you can hold about a person, and the predictions you can make about them, and those limits are derived from the complete set of your previous experiences of that person, leaving out nothing.

Another way to say this is that, in my culture, vague suspicions and unfounded intuitions are allowed only insofar as they are openly flagged as such, and in general will not carry much weight when they're in conflict with a consistent, coherent, observable record.

(Although if a vague suspicion or unfounded intuition is born out, then the person who registered it will have earned a solid point in favor of being *good at making accurate predictions that run counter to existing data*).

This also means that updates about a person must be gradual and Bayesian, and that in general people are expected to seek the smallest possible update. If a seemingly loyal colleague has seemingly betrayed you, you are obligated in my culture to be confused and curious and look for alternate explanations, rather than leaping straight to a new story which contradicts all of your prior evidence. If you learn an unpleasant secret about someone (e.g. that they privately abuse heroin), your update is confined to the domain of that secret, and can't be used to invalidate or dismiss their record of e.g. completing work projects on time or being kind to animals and the elderly.

(You might easily still be justified in firing them, if they are your employee, or ceasing to associate with them, if they are your friend. For instance, the new data might actually be relevant to your estimate of entangled qualities like their integrity or conscientiousness or decisionmaking or trustworthiness. But in general, even if the new data is damning, people in my culture make a conscious effort to compensate for the human impulse to round everything off to a single answer.[3] Sometimes people aren't simple or straightforward, and in my culture you can object to the costs associated with one trait without unjustly and inaccurately lowering your evaluation of the rest).

Symmetrically, in my culture, there are limits to the kinds of claims you can make *about yourself.* If someone does not believe that you have Trait X, and you believe that you do, you're not allowed to just *declare* that you do, and expect them to update. Instead, the most you can do is say something like "Look, of course you're going to hold this with some skepticism, I endorse that—not that you *needed* me to endorse it, it's not like you need my permission—but I claim I have Trait X, and I'd like you to accept that as a hypothesis, and give me a chance to prove it. Like, I'm requesting that you keep your radar open and be receptive to evidence on this question".

• • •

In my culture, unless you've made a specific commitment to the contrary, you can end any conversation at any time.

There are ways to do this which might be blunt or rude or cruel, and people might dock you social points for your bluntness or rudeness or cruelty.

But they will not dock you points *for the leaving itself.* In my culture, you do not owe anyone your attention, your ear, or your thoughts. If you and I were romantically engaged, and we broke up, and I am desperate for closure, that is *my* problem, not yours. If you *want* to provide me the opportunity to talk further, that's great, but I have no right to demand it, and if I'm being pushy, other people will step in to remind you that you're morally justified in simply cutting me off.

• • •

In my culture, if I cover the check and you agree to PayPal me your share, it is not up to me to remind you.

In my culture, if I lend you a book, I should not have to ask for it back.

In my culture, if you break something that I own, it goes without saying that you will take the initiative to replace it, or reimburse me, or make it up to me in some other fashion.

In my culture, failing to fulfill an obligation until someone else reminds you to do so is itself viewed as a transgression. It's seen as imposing an additional cost upon your creditor, and onlookers will tend to notice that, and remember. There's an *instrumental* argument for people to remind you anyway (since otherwise they might never get what you owe them), but in that case, it's understood that you've lost an additional point, and have to either accept that fact or make up for it with some kind of concrete action.

• • •

In my culture, if you're going to a Cirque du Soleil show, and one of your friends is a gymnast, it is viewed as slightly rude to fail to think of inviting them. Ditto if you're planning a night of Magic: The Gathering and one of your friends is always lamenting a lack of people to play with.

You don't have to *actually* invite them, if there's reason not to. For instance, if you simply wouldn't enjoy their company, or if they're incompatible with someone else who's already going. But it's expected that you will *pause to ask yourself the question.* That you will *notice the opportunity* to include them, especially if their interest is sufficiently well-known that they might reasonably *expect* their name to come to mind. To fail to do so twice in a row will undoubtedly be taken as a signal that you dislike them or don't particularly care for them or are willing to prioritize other things above their preferences in this domain, and they will update their expectations with regards to you accordingly.

(There's an interpretation of this which sounds vaguely threatening, à la "invite me or else". But in my culture it's actually much gentler and more stoic than that—it's more like "invite me, or I'll update toward believing that you're not the kind of person who will invite me." That doesn't have a strong moral shading of any kind, in my culture—it's just *using the past to make predictions about the future*).

• • •

In my culture, it is seen as trivially obvious that if a friend has to *ask* you whether or not you want a hug, the friendship is weak enough that the hug will probably not be particularly nourishing in the first place (since it comes with a reminder of distance and insecurity).

This is not to say that all hugs are always welcome—it's a bad idea to simply hug people willy-nilly, or to offer hugs as a way to cargo-cult your way into a closer relationship. If you're not sure, you ask anyway and pay the distance cost, or you just don't hug at all.

Rather, the thing you aim for is to grow your friendship to the point where your friend knows you well enough, and feels secure enough in their relationship with you, that they can *tell* when you need a hug, and aren't the slightest bit anxious that it'll be taken badly if they happen to be wrong.

• • •

In my culture, birthdays matter, and celebrating birthdays matters, and remembering to celebrate birthdays matters, and actually knowing what kind of celebration will be warm and wanted is more important and more highly valued than the mere desire to do so.

(Another way to say this is that, in my culture, the thought *does* count, but other things count *more*).

• • •

In my culture, we play punch bug.[4]

• • •

In my culture, if I'm trying to help you, it's important both that it's not about me and that it doesn't *seem*, on the surface, that it's about me.

(Note the "if". Sometimes, the situation can be more complex, and my primary goal is not simply to help you, but to investigate a situation, or to protect myself along with you, or to preserve community norms and standards, etc).

But *if* I'm actually in it for your benefit, as opposed to because your feeling bad makes me feel bad and what I really want is to stop feeling bad *myself*—

If that is the case, then it's important to avoid setting up a dynamic whereby you, who are already struggling and low on resources, feel pressure to reassure *me*. I don't want to make it seem like my own failure to help will mean that I will not be okay, such that you have to put on a mask and *pretend to have been helped*, regardless of whether or not you really have. In my culture, we recognize that this is the opposite of helping—it's adding yet another item onto the pile of stuff you have to deal with.

(Which is not to say that I'm not allowed to *feel distress*, and possibly take actions in accordance with that distress, when someone else is sending up distressing signals. If you express extreme depression and suicidal impulses to me, for instance, *of course* I'm going to have an emotional reaction).

But absent any of those other considerations—if it's just you and me in a room and my goal is to help—if I want to have any chance of actually helping you, it's understood in my culture that it has to be clear that my own ego is not relevantly in the mix. It has to be clear that I'm *okay with you being sad/mad/stressed.* It has to be clear that I'm there to offer you tools and support to get at the root cause, not to punish you for your symptoms, and furthermore that if you don't want my tools and support, or don't find them helpful in the end, that's okay too.

• • •

In my culture, we distinguish between *what a situation looks like*[5] *and what it actually is.*

When we can't tell the difference, we intentionally form two reactions—one appropriate for each possibility—and we take only provisional and reversible actions (or actions that are good in *both* branches of possibility) until we have some hint as to what's actually true.

(To make this more concrete: in my culture, there simply isn't anything like the Duke lacrosse scandal.[6] That sort of unjustified public excoriation just doesn't happen).

• • •

In my culture, we believe that a peace treaty is not a suicide pact.

Civilization works (at least in part) by mutual agreement to *set aside options*. As the Freud quote goes, "the first person to hurl an insult instead of a stone was the founder of civilization". We learn in kindergarten not to hit each other as a method of expressing disagreement—we *mutually sacrifice* that option, trading the ability to hit at will for the safety of not having to be constantly on alert against being hit.

But in my culture, there's a distinction between "I agree to follow these rules with everyone I interact with" and "I agree to follow these rules with everyone who also agrees with them". This can be tricky when it comes to balancing my culture with the needs of the broader context culture that allows us all to coexist. If Lord Duellington comes from a culture where people get into duels over perceived insults, it needs to be *really crystal clear* to everyone else at the diplomatic summit that Lord Duellington *really understands that this aspect of his culture is not shared*, and that there is zero chance that Lord Duellington will propose a duel, regardless of how insulted he perceives himself to be.

Even the mere *perception* of Lord Duellington as a loose cannon who cannot distinguish between [his own local culture] and [what's agreed-upon by everyone present] is costly, regardless of whether or not it's true, because it results in a dynamic where other people at the summit feel like they have to spend lots of energy and attention tracking and maintaining their personal physical safety, rather than being able to rely on the social contract. Thus, it's important that Lord Duellington not even mutter under his breath that *back home, we have a way of dealing with people like you*, and it's entirely reasonable for people to disinvite Lord Duellington from future summits if he does.

(Although it's also reasonable in my culture for Lord Duellington to object that his track record indicates that he would never actually do such a thing, and that it's not fair for people to see him as a risk, as noted above. But it can *both* be true that he's not actually a direct risk, and *also* be true that his presence nevertheless puts too much stress on the social fabric to be worth it, and even Lord Duellington will admit that this is possible).

This is why, for instance, I no longer use my car's horn as a tool for reacting to perceived defections in traffic. In my home culture, certain traffic violations effectively equate to *opting out* of the treaty whereby people don't honk at each other.

In the Bay Area context culture, though, that's a sort of violence that isn't justified *even by overt defection*, as far as I can tell. People will socially punish it even if they agree that the honker was wronged, because use of the horn for that purpose is seen as corrosive in other ways.

(For instance, it hurts nearby people as well, and reduces the effectiveness of the horn as an exclusively emergency signal, and grabs people in a sort of deep, visceral, monkey-brain kind of way that is almost guaranteed to escalate the situation rather than resolving it).

And it seems correct to conform to the context culture in this way, and for others to expect *you* to conform, regardless of your home culture. A clause that releases you from the social contract *in your home culture* does not have the power to release you from the social contract *at a diplomatic summit.* The whole *point* of a diplomatic summit is that people are setting aside their native standards for what constitutes transgression, and agreeing to a cooperative compromise culture. If a person isn't willing to make that sacrifice, then the diplomatic summit can't afford to have them around, no matter what value is lost in the process of excluding them.

But in my *home* culture, if Cameron is behaving toward Dallas in ways that are proscribed by social norms, and Dallas has tried a couple of times to get Cameron to stop, and has actually made efforts to bridge the gap, and has taken actions which reasonably rule out the possibility that Cameron was just making a one-time mistake out of frustration or misunderstanding or whatever—

In my home culture, if Dallas retaliates in kind, no one will assign Dallas the sort of blame they would otherwise receive for taking a forbidden action.

And if Elliott steps in to protect Dallas from Cameron, and uses Cameron's tactics against them, no one will look sideways at Elliott, either.

In my culture, there's a fundamental difference between people who are *trying to play by the rules,* and people who are *unmistakably willing to violate them.* The former category gets a lot of protection, and the latter gets meaningfully less, especially when it comes to open antagonism. In my culture, we try to be really damn sure that it's *actually* fire, and not just smoke, but we are indeed willing to fight fire with fire, and we don't dock one another points for doing so.

I sometimes wish that other specific subcultures had more in common with my home culture in this way,[7] but I think that it's correct that the broader context culture does not, because the whole thing rests on the idea that there can be clear common knowledge about what constitutes a transgression and what doesn't, and that everyone can see what's justified and what isn't, and that seems to me to be an impossibly naive expectation when we're talking about a diplomatic summit with representatives from thousands of individual cultures.

Where to go from here

This is not at all a complete list. On reflection, it's probably not even the *most* important differences between my culture and the general culture—just some of the more important ones, the ones which floated to the top over the two weeks I spent tinkering with this essay. No doubt I'll discover more after hitting 'publish.' But it's okay that this is incomplete, because me sharing *my culture* is only half of the process. The other half is you sharing *yours*, and so this feels like as good a place as any to pause and listen, in case you feel like talking.

Originally published 7th March 2019

(1) *Wikipedia. Nonviolent Communication.*

(2) *Chris Leong. Decoupling vs Contextualising Norms. 14th May 2018.*

(3) *Nagesh Belludi. The Halo and Horns Effects [Rating Errors]. 30th Apr 2010.*

(4) *Duncan A Sabien. In Defense of Punch Bug. 3rd May 2018.*

(5) *Duncan A Sabien. It's Not What It Looks Like. 5th Aug 2018.*

(6) *Wikipedia. Duke lacrosse case.*

(7) *Duncan A Sabien. Moderating LessWrong. 24th May 2018.*

Evan Hubinger

Chris Olah's views on AGI safety

Note: I am not Chris Olah. This post was the result of lots of back-and-forth with Chris, but everything here is my interpretation of what Chris believes, not necessarily what he actually believes. Chris also wanted me to emphasize that his thinking is informed by all of his colleagues on the OpenAI Clarity team and at other organizations.

In thinking about AGI safety—and really any complex topic on which many smart people disagree—I've often found it very useful to build a collection of different viewpoints from people that I respect that I feel like I understand well enough to be able to think from their perspective. For example, I will often try to compare what an idea feels like when I put on my Paul Christiano hat to what it feels like when I put on my Scott Garrabrant hat. Recently, I feel like I've gained a new hat that I've found extremely valuable that I also don't think many other people in this community have, which is my Chris Olah hat. The goal of this post is to try to give that hat to more people.

If you're not familiar with him, Chris Olah leads the Clarity team at OpenAI and formerly used to work at Google Brain. Chris has been a part of many of the most exciting ML interpretability results in the last five years, including Activation Atlases,[1] Building Blocks of Interpretability,[2] Feature Visualization,[3] and DeepDream.[4] Chris was also a coauthor of "Concrete Problems in AI Safety".[5]

He also thinks a lot about technical AGI safety and has a lot of thoughts on how ML interpretability work can play into that—thoughts which, unfortunately, haven't really been recorded previously. So: here's my take on Chris's AGI safety worldview.

The benefits of transparency and interpretability

Since Chris primarily works on ML transparency and interpretability, the obvious first question to ask is how he imagines that sort of research aiding with AGI safety. When I was talking with him, Chris listed four distinct ways in which he thought transparency and interpretability could help, which I'll go over in his order of importance.

Catching problems with auditing

First, Chris says, interpretability gives you a mulligan. Before you deploy your AI, you can throw all of your interpretability tools at it to check and see what it actually learned and make sure it learned the right thing. If it didn't—if you find that it's learned some sort of potentially dangerous proxy, for example—then you can throw your AI out and try again. As long as you're in a domain where your AI isn't actively trying to deceive your interpretability tools (via deceptive alignment,[6] perhaps), this sort of a mulligan could help quite a lot in resolving more standard robustness problems (proxy alignment,[7] for example). That being said, that doesn't necessarily mean waiting until you're on the verge of deployment to look for flaws. Ideally you'd be able to discover problems early on via an ongoing auditing process as you build more and more capable systems.

One of the OpenAI Clarity team's major research thrusts right now is developing the ability to more rigorously and systematically audit neural networks. The idea is that interpretability techniques shouldn't have to "get lucky" to stumble across a problem, but should instead reliably catch any problematic behavior. In particular, one way in which they've been evaluating progress on this is the "auditing game". In the auditing game, one researcher takes a neural network and makes some modification to it—maybe images containing both dogs and cats are now classified as rifles, for example—and another researcher, given only the modified network, has to diagnose the problem and figure out exactly what modification was made to the network using only interpretability tools without looking at error cases. Chris's hope is that if we can reliably catch problems in an adversarial context like the auditing game, it'll translate into more reliably being able to catch alignment issues in the future.

Deliberate design

Second, Chris argues, advances in transparency and interpretability could allow us to significantly change the way we design ML systems. Instead of a sort of trial-and-error process where we just throw lots of different techniques at the various benchmarks and see what sticks, if we had significantly better transparency tools we might be able to design our systems deliberately by understanding why our models work and how to improve them. In this world, because we would be building systems with an understanding of why they work, we might be able to get a much better understanding of their failure cases as well and how to avoid them.

In addition to these direct benefits, Chris expects some large but harder-to-see benefits from such a shift as well. Right now, not knowing anything about how your model works internally is completely normal. If even partly understanding one's model became normal, however, then the amount we don't know might become glaring and concerning. Chris provides the following analogy to illustrate this: if the only way you've seen a bridge be built before is through unprincipled piling of wood, you might not realize what there is to worry about in building bigger bridges. On the other hand, once you've seen an example of carefully analyzing the structural properties of bridges, the absence of such an analysis would stand out.

Giving feedback on process

Third, access to good transparency and interpretability tools lets you give feedback to a model—in the form of a loss penalty, reward function, etc.—not just on its output, but also on the process it used to get to that output. Chris and his coauthors lay this argument out in "Building Blocks of Interpretability":[8]

> One very promising approach to training models for these subtle objectives is learning from human feedback. However, even with human feedback, it may still be hard to train models to behave the way we want if the problematic aspect of the model doesn't surface strongly in the training regime where humans are giving feedback. Human feedback on the model's decision-making process, facilitated by interpretability interfaces, could be a powerful solution to these problems. It might allow us to train models not just to make the right decisions, but to make them for the right reasons. (There is however a danger here: we are optimizing our model to look the way we want in our interface—if we aren't careful, this may lead to the model fooling us)!

The basic idea here is that rather than just using interpretability as a mulligan at the end, you could also use it as part of your objective during training, incentivizing the model to be as transparent as possible. Chris notes that this sort of thing is quite similar to the way in which we actually judge human students by asking them to show their work. Of course, this has risks—it could increase the probability that your model only looks transparent but isn't actually—but it also has the huge benefit of helping your training process steer clear of bad uninterpretable models. In particular, I see this as potentially being a big boon for informed oversight,[9] as it allows you to incorporate into your objective an incentive to be more transparent to an amplified overseer.

One way in particular that the Clarity team's work could be relevant here is a research direction they're working on called model diffing. The idea of model diffing is to have a way of systematically comparing different models and determining what's different from the point of view of high-level concepts and abstractions. In the context of informed oversight—or specifically relaxed adversarial training[10]—you could use model diffing to track exactly how your model is evolving over the course of training in a way which is inspectable to the overseer.[1]

1. In particular, this could be a way of getting traction on addressing gradient hacking.

Building microscopes not agents

One point that Chris likes to talk about is that—despite talking a lot about how we want to avoid race-to-the-bottom dynamics—the AI safety community seems to have just accepted that we have to build agents, despite the dangers of agentic AIs.[2] Of course, there's a reason for this: agents seem to be more competitive. Chris cites Gwern's "Why Tool AIs Want to Be Agent AIs"[11] here, and notes that he mostly agrees with it—it does seem like agents will be more competitive, at least by default.

2. As an example of the potential dangers of agents, more agentic AI setups seem much more prone to mesa-optimization.

But that still doesn't mean we have to build agents—there's no universal law compelling us to do so. Rather, agents only seem to be on the default path because a lot of the people who currently think about AGI see them as the shortest path.[3] But potentially, if transparency tools could be made significantly better, or if a major realignment of the ML community could be achieved—which Chris thinks might be possible, as I'll talk about later—then there might be another path.

3. A notable exception to this, however, is Eric Drexler's "Reframing Superintelligence: Comprehensive AI Services as General Intelligence".

Specifically, rather than using machine learning to build agents which directly take actions in the world, we could use ML as a *microscope*—a way of learning about the world without directly taking actions in it. That is, rather than training an RL agent, you could train a predictive model on a bunch of data and use interpretability tools to inspect it and figure out what it learned, then use those insights to inform—either with a human in the loop or in some automated way—whatever actions you actually want to take in the world.

Chris calls this alternative vision of what an advanced AI system might look like a microscope AI since the AI is being used sort of like a microscope to learn about and build models of the world. In contrast with something like a tool or oracle AI that is designed to output useful information, the utility of a microscope AI wouldn't come from its output but rather our ability to look inside of it and access all of the implicit knowledge it learned. Chris likes to explain this distinction by contrasting Google Translate—the oracle/tool AI in this analogy—to an interface that could give you access to all the linguistic knowledge implicitly present in Google Translate—the microscope AI.

Chris talks about this vision in his post "Visualizing Representations: Deep Learning and Human Beings":[12]

> The visualizations are a bit like looking through a telescope. Just like a telescope transforms the sky into something we can see, the neural network transforms the data into a more accessible form. One learns about the telescope by observing how it magnifies the night sky, but the really remarkable thing is what one learns about the stars. Similarly, visualizing representations teaches us about neural networks, but it teaches us just as much, perhaps more, about the data itself.
>
> (If the telescope is doing a good job, it fades from the consciousness of the person looking through it. But if there's a scratch on one of the telescope's lenses, the scratch is highly visible.

> If one has an example of a better telescope, the flaws in the worse one will suddenly stand out. Similarly, most of what we learn about neural networks from representations is in unexpected behavior, or by comparing representations).
>
> Understanding data and understanding models that work on that data are intimately linked. In fact, I think that understanding your model has to imply understanding the data it works on.
>
> While the idea that we should try to visualize neural networks has existed in our community for a while, this converse idea—that we can use neural networks for visualization—seems equally important [and] is almost entirely unexplored.

Shan Carter and Michael Nielsen have also discussed similar ideas in their Artificial Intelligence Augmentation[13] article in Distill.

Of course, the obvious question with all of this is whether it could ever be anything but hopelessly uncompetitive. It is important to note that Chris generally agrees that microscopes are unlikely to be competitive—which is why he's mostly betting on the other routes to impact above. He just hasn't entirely given up hope that a realignment of the ML community away from agents towards things like deliberate design and microscopes might still be possible.

Furthermore, even in a world where the ML community still looks very similar to how it does today, if we have really good interpretability tools and the largest AI coalition has a strong lead over the next largest, then it might be possible to stick with microscopes for quite some time. Perhaps enough to either figure out how to align agents or otherwise get some sort of decisive strategic advantage.

What if interpretability breaks down as AI gets more powerful?

Chris notes that one of the biggest differences between him and many of the other people in the AI safety community is his belief that very strong interpretability is at all possible. The model that Chris has here is something like a reverse compilation process that turns a neural network into human-understable code. Chris notes that the resulting code might be truly gigantic—e.g. the entire Linux kernel—but that it would be faithful to the model and understandable by humans. Chris's basic intuition here is that neural networks really do seem to learn meaningful features and that if you're willing to put a lot of energy in to understand them all—e.g. just actually inspect every single neuron—then you can make it happen. Chris notes that this is in contrast to a lot of other neural network interpretability work which is more aimed at approximating what neural networks do in particular cases.

Of course, this is still heavily dependent on exactly what the scaling laws are like for how hard interpretability will be as our models get stronger and more sophisticated. Chris likes to use the following graph to describe how he sees transparency and interpretability tools scaling up:

This diagram tries to capture intuition, not any formal / precise truth.

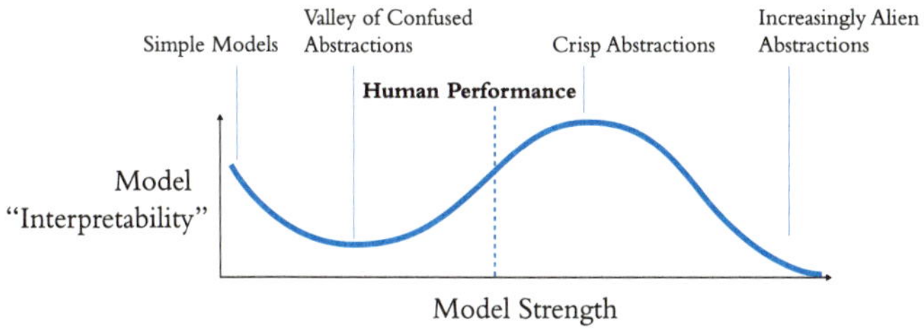

This graph has a couple of different components to it. First, simple models tend to be pretty interpretable—think for example linear regression, which gives you super easy-to-understand coefficients. Second, as you scale up past simple stuff like linear regression, things get a lot messier. But Chris has a theory here: the reason these models aren't very interpretable is because they don't have the capacity to express the full concepts that they need, so they rely on confused concepts that don't quite track the real thing. In particular, Chris notes that he has found that better, more advanced, more powerful models tend to have crisper, clearer, more interpretable concepts—e.g. InceptionV1 is more interpretable than AlexNet. Chris believes that this sort of scaling up of interpretability will continue for a while until you get to around human-level performance, at which point Chris hypothesizes that the trend will stop as models start moving away from crisp human-level concepts to still crisp but now quite alien concepts.

If you buy this graph—or something like it—then interpretability should be pretty useful all the way up to and including AGI—though perhaps not for very far past AGI. But if you buy a continuous-takeoff worldview, then that's still pretty useful. Furthermore, in my opinion, I think that the dropping off of interpretability at the end of this graph is just an artifact of using a human overseer. If you instead substituted in an amplified overseer, then I think it's plausible that interpretability could just keep going up, or at least level off at some high level.

Improving the field of machine learning

One thing that Chris thinks could really make a big difference in achieving a lot of the above goals would be some sort of realignment of the machine learning community. Currently, the thing that the ML community primarily cares about is chasing state-of-the-art results on its various benchmarks without regard for understanding what the ML tools they're using are actually doing. But that's not what the machine learning disciplines has to look like, and in fact, it's not what most scientific disciplines do look like.

Here's Chris's vision for what an alternative field of machine learning might look like. Currently, machine learning researchers primarily make progress on benchmarks via trial and error. Instead, Chris wants to see a field which focuses on deliberate design where understanding models is prioritized and the way that people make progress is through deeply understanding their systems. In this world, ML researchers primarily make better models by using interpretability tools to understand why their models are doing what they're doing instead of just throwing lots of things at the wall and seeing what sticks. Furthermore, a large portion of the field in this world is just devoted to gathering information on what models do—cataloging all the different types of circuits that appear across different neural net-

4. An example of the sort of common circuit that appears in lots of different models that the Clarity team has found is the way in which convolutional neural networks stay reflection-invariant: to detect a dog, they separately detect leftwards-facing and rightwards-facing dogs and then union them together.

5. This results in a large portion of the field being focused on what is effectively microscope use, which could also be quite relevant for making microscope AIs more tenable.

works, for example[4]—rather than on trying to build new models.[5]

If you want to change the field in this way, there are essentially two basic paths to making something like that happen—you can either:

1. get current ML researchers to switch over to interpretability/ deliberate design/microscope use or
2. produce new ML researchers working on those things.

Chris has thoughts on how to do both of these, but I'll start with the first one. Chris thinks that several factors could make a high-quality interpretability field appealing for researchers. First, interpretability could be a way for researchers without access to large amounts of compute to stay relevant in a world where relatively few labs can train the largest machine learning models. Second, Chris thinks there's lots of low hanging fruit in interpretability such that it should be fairly easy to have impressive research results in the space over the next few years. Third, Chris's vision of interpretability is very aligned with traditional scientific virtues—which can be quite motivating for many people—even if it isn't very aligned with the present paradigm of machine learning.

However, If you want researchers to switch to a new research agenda and/or style of research, it needs to be possible for them to support careers based on it. Unfortunately, the unit of academic credit in machine learning tends to be traditional papers, published in conferences, evaluated on whether they set a new state-of-the-art on a benchmark (or more rarely by proving theoretical results). This is what decides who gets hired, promoted, and tenured in machine learning.

To address this, Chris founded Distill,[14] an academic machine learning journal that aims to promote a different style of machine learning research. Distill aims to be a sort of "adapter" between the traditional method of evaluating research and the new style of research—based around things like deliberate design and microscope use—that Chris wants to see the field move to. Specifically, Distill does this by being different in a few key ways:

1. Distill explicitly publishes papers visualizing machine learning systems, or even just explanations improving Clarity of thought in machine learning (Distill's expository articles have become widely used references).

2. Distill has all of the necessary trappings to make it recognized as a legitimate academic journal such that Distill publications will be taken seriously and cited.[15]

3. Distill has support for all the sorts of nice interactive diagrams that are often necessary for presenting interpretability research. The second option is to produce new ML researchers pursuing deliberate design rather than converting old ones. Here, Chris has a pretty interesting take on how this can be done: convert neuroscientists and systems biologists.

Here's Chris's pitch. There are whole fields of neuroscience dedicated to understanding all the different connections, circuits, pathways, etc. in all different manner of animal brains. Similarly, for the systems biologists, there are significant communities of researchers studying individual proteins, their interactions and pathways, etc. While neural networks are different from these lines of research at a detailed level, a lot of high level research expertise—e.g. epistemic standards for studying circuits, recurring motifs, research intuition—may be just as helpful for this type of research as machine learning expertise. Chris thinks neuroscientists or systems biologists willing to make this transition would be able to get funding to do their research, a much easier time running experiments, and lots of low-hanging fruit in terms of new publishable results that nobody has found yet.

Doesn't this speed up capabilities?

Yes, it probably does—and Chris agrees that there's a negative component to that—but he's willing to bet that the positives outweigh the negatives.

Specifically, Chris thinks the main question is whether principled and deliberate model design based on interpretability can beat automated model design approaches like neural architecture search. If it can, we get capabilities acceleration, but also a paradigm shift towards deliberate model design, which Chris expects to significantly aid alignment. If we don't, interpretability loses one of its upsides (other advantages like auditing still exist in this world) but also doesn't have the downside of acceleration. Both the upside and downside go hand in hand, and Chris expects the upside to outweigh the downside.

• • •

Update: If you're interested in understanding Chris's current transparency and interpretability work, a good starting point is the Circuits Thread[16] *on Distill.*

Comment from Chris Olah

Evan, thank you for writing this up! I think this is a pretty accurate description of my present views, and I really appreciate you taking the time to capture and distill them. :)

One thing I wanted to emphasize is that, to the extent these views seem intellectually novel to members of the alignment community, I think it's more accurate to attribute the novelty to a separate intellectual community loosely clustered around Distill than to me specifically. My views are deeply informed by the thinking of other members of the Clarity team and our friends at other institutions. To give just one example, the idea presented here as a "microscope AI" is deeply influenced by Shan Carter and Michael Nielsen's thinking, and the actual term was coined by Nick Cammarata.

To be clear, not everyone in this community would agree with my views, especially as they relate to safety and strategic considerations! So I shouldn't be taken as speaking on behalf of this cluster, but rather as articulating a single point of view within it.

(1) *Shan Carter, Zan Armstrong, Ludwig Schubert, Ian Johnson, Chris Olah. Exploring Neural Networks with Activation Atlases. 6th Mar 2019.*

(2) *Chris Olah, Arvind Satyanarayan, Ian Johnson, Shan Carter, Ludwig Schubert, Katherine Ye, Alexander Mordvintsev. The Building Blocks of Interpretability. 6th Mar 2018.*

(3) *Chris Olah, Alexander Mordvintsev, Ludwig Schubert. Feature Visualization. 7th Nov 2017.*

(4) *Alexander Mordvintsev, Software Engineer, Christopher Olah, Software Engineering Intern and Mike Tyka, Software Engineer. Inceptionism: Going Deeper into Neural Networks. 17th Jun 2015.*

(5) *Dario Amodei, Chris Olah, Jacob Steinhardt, Paul Christiano, John Schulman, Dan Mané. Concrete Problems in AI Safety. 25th Jul 2016.*

(6) *Evan Hubinger, Chris van Merwijk, vlad_m, Joar Skalse, Scott Garrabrant. Deceptive Alignment. 5th Jun 2019.*

(7) *Evan Hubinger, Chris van Merwijk, vlad_m, Joar Skalse, Scott Garrabrant. The Inner Alignment Problem. 3rd Jun 2019.*

(8) *Christopher Olah, Arvind Satyanarayan, Ian Johnson, Shan Carter, Ludwig Schubert, Katherine Ye, Alexander Mordvintsev. The Building Blocks of Interpretability. 6th Mar 2018.*

(9) *Paul Christiano. Informed oversight. 9th Jan 2019.*

(10) *Evan Hubinger. Relaxed adversarial training for inner alignment. 10th Sep 2019.*

(11) *Gwern.net. Why Tool AIs Want to Be Agent AIs. 28th Aug 2018.*

(12) *Christopher Olah. Visualizing Representations: Deep Learning and Human Beings. 16th Jan 2015.*

(13) *Shan Carter, Michael Nielsen. Using Artificial Intelligence to Augment Human Intelligence. 4th Dec 2017.*

(14) *Distill. About.*

(15) *Google Scholar, Search. "site:distill.pub" (https://scholar.google.com/scholar?q=site%3Adistill.pub)*

(16) *Thread: Circuits. 10th Mar 2020.*

Understanding "Deep Double Descent"

If you're not familiar with the double descent phenomenon, I think you should be. I consider double descent to be one of the most interesting and surprising recent results in analyzing and understanding modern machine learning. Today, Preetum et al. released a new paper, "Deep Double Descent",[1] which I think is a big further advancement in our understanding of this phenomenon. I'd highly recommend at least reading the summary of the paper on the OpenAI blog.[2] However, I will also try to summarize the paper here, as well as give a history of the literature on double descent and some of my personal thoughts.

Prior work

The double descent phenomenon was first discovered by Mikhail Belkin et al., who were confused by the phenomenon wherein modern ML practitioners would claim that "bigger models are always better" despite standard statistical machine learning theory predicting that bigger models should be more prone to overfitting. Belkin et al. discovered that the standard bias-variance tradeoff picture actually breaks down once you hit approximately zero training error—what Belkin et al.[3] call the "interpolation threshold." Before the interpolation threshold, the bias-variance tradeoff holds and increasing model complexity leads to overfitting, increasing test error. After the interpolation threshold, however, they found that test error actually starts to go down as you keep increasing model complexity! Belkin et al. demonstrated this phenomenon in simple ML methods such as decision trees as well as simple neural networks trained on MNIST. Here's the diagram that Belkin et al. use in their paper to describe this phenomenon:

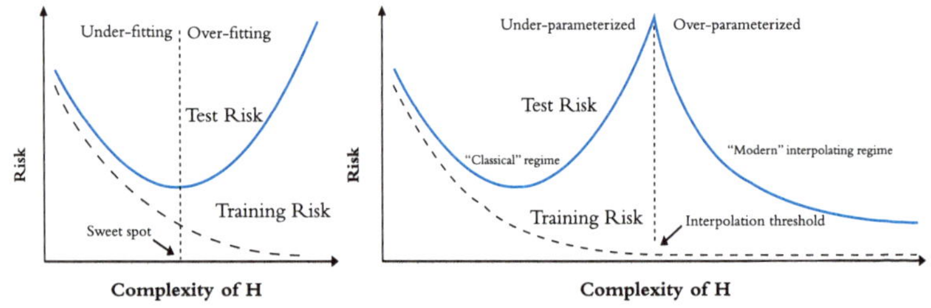

(a) U-shaped "bias-variance" risk curve (b) "double descent" risk curve

Belkin et al. describe their hypothesis for what's happening as follows:

> All of the learned predictors to the right of the interpolation threshold fit the training data perfectly and have zero empirical risk. So why should some—in particular, those from richer

> functions classes—have lower test risk than others? The answer is that the capacity of the function class does not necessarily reflect how well the predictor matches the inductive bias appropriate for the problem at hand. [The inductive bias] is a form of Occam's razor: the simplest explanation compatible with the observations should be preferred. By considering larger function classes, which contain more candidate predictors compatible with the data, we are able to find interpolating functions that [are] "simpler". Thus increasing function class capacity improves performance of classifiers.

I think that what this is saying is pretty magical: in the case of neural nets, it's saying that SGD just so happens to have the right inductive biases that letting SGD choose which model it wants the most out of a large class of models with *the same training performance* yields significantly better test performance. If you're right on the interpolation threshold, you're effectively "forcing" SGD to choose from a very small set of models with perfect training accuracy (maybe only one realistic option), thus ignoring SGD's inductive biases completely—whereas if you're past the interpolation threshold, you're letting SGD choose which of many models with perfect training accuracy it prefers, thus allowing SGD's inductive bias to shine through.

I think this is strong evidence for the critical importance of implicit simplicity and speed priors in making modern ML work. However, such biases also produce strong incentives for mesa-optimization[4] (since optimizers are simple, compressed policies) and pseudo-alignment[5] (since simplicity and speed penalties will favor simpler, faster proxies). Furthermore, the arguments for the universal prior[6] and minimal circuits[7] being malign suggest that such strong simplicity and speed priors could also produce an incentive for deceptive alignment.[8]

"Deep Double Descent"

Now we get to Preetum et al.'s new paper, "Deep Double Descent". Here are just some of the things that Preetum et al. demonstrate in "Deep Double Descent":

1. double descent occurs across a wide variety of different model classes, including ResNets, standard CNNs, and Transformers, as well as a wide variety of different tasks, including image classification and language translation,
2. double descent occurs not just as a function of model size, but also as a function of *training time* and *dataset size*, and
3. since double descent can happen as a function of dataset size, **more data can lead to worse test performance!**

Crazy stuff. Let's try to walk through each of these results in detail and understand what's happening.

First, double descent is a highly universal phenomenon in modern deep learning. Here is double descent happening for ResNet18 on CIFAR-10 and CIFAR-100:

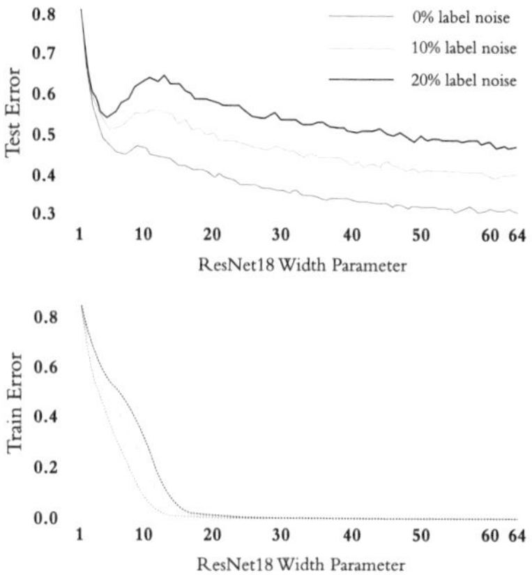

(a) **CIFAR-100**. There is a peak in test error even with no label noise.

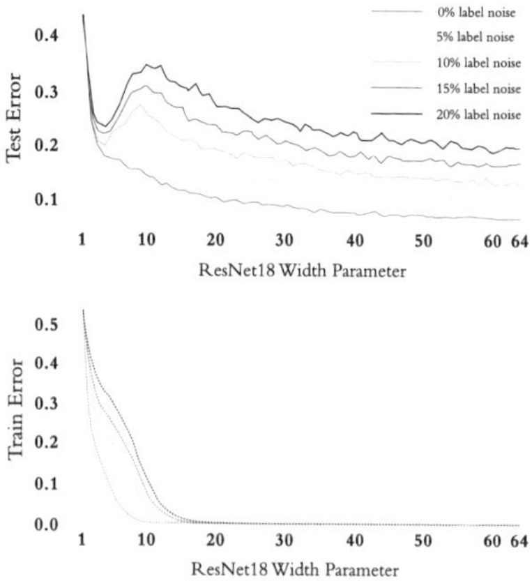

(a) **CIFAR-10**. There is a "plateau" in test error around the interpolation point with no label noise, which develops into a peak for added label noise.

And again for a Transformer model on German-to-English and English-to-French translation:

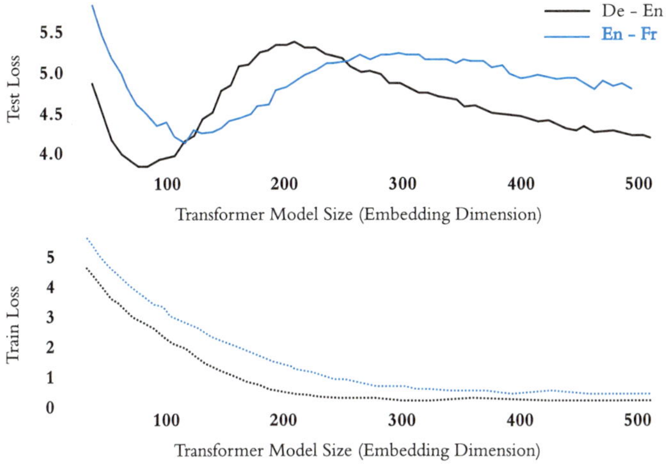

All of these graphs, however, are just showcasing the standard Belkin et al.-style double descent over model size (what Preetum et al. call "model-wise double descent"). What's really interesting about "Deep Double Descent," however, is that Preetum et al. also demonstrate that the same thing can happen for training time ("epoch-wise double descent") and a similar thing for dataset size ("sample-wise non-monotonicity").

First, let's look at epoch-wise double descent. Take a look at these graphs for ResNet18 on CIFAR-10:

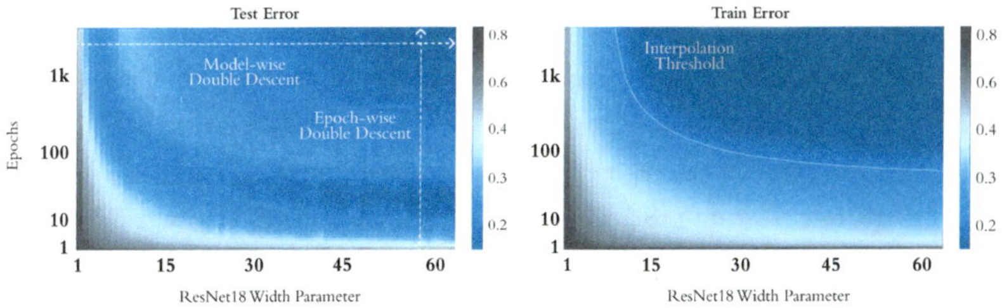

There's a bunch of crazy things happening here which are worth pointing out. First, the obvious: epoch-wise double descent is definitely a thing—holding model size fixed and training for longer exhibits the standard double descent behavior. Furthermore, the peak happens right at the interpolation threshold where you hit zero training error. Second, notice where you don't get epoch-wise double descent: if your model is too small to ever hit the interpolation threshold—like was the case in ye olden days of ML—you never get epoch-wise double descent. Third, notice the log scale on the y axis: you have to train for quite a while to start seeing this phenomenon.

Finally, sample-wise non-monotonicity—Preetum et al. find a regime where increasing the amount of training data by *four and a half times* actually *increases* test loss (!):

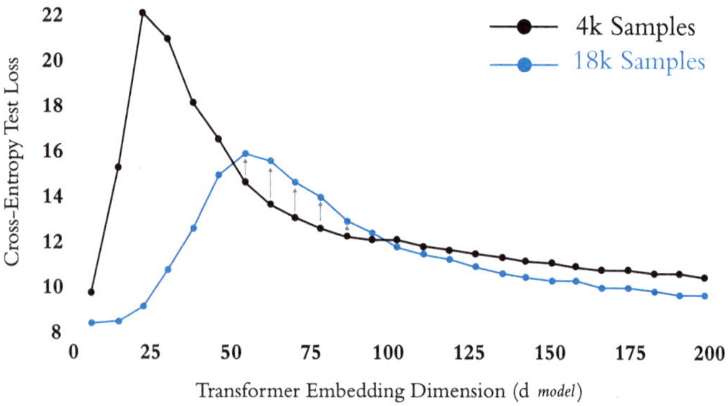

What's happening here is that more data increases the amount of model capacity/number of training epochs necessary to reach zero training error, which pushes out the interpolation threshold such that you can regress from the modern (interpolation) regime back into the classical (bias-variance tradeoff) regime, decreasing performance.

Additionally, another thing which Preetum et al. point out which I think is worth talking about here is the impact of label noise. Preetum et al. find that increasing label noise significantly exaggerates the test error peak around the interpolation threshold. Why might this be the case? Well, if we think about the inductive biases story from earlier, greater label noise means that near the interpolation threshold SGD is forced to find the one model which fits all of the noise—which is likely to be pretty bad since it has to model a bunch of noise. After the interpolation threshold, however, SGD is able to pick between many models which fit the noise and select one that does so in the simplest way such that you get good test performance.

Final comments

I'm quite excited about "Deep Double Descent", but it still leaves what is in my opinion the most important question unanswered, which is: what exactly are the magical inductive biases of modern ML that make interpolation work so well?

One proposal I am aware of is the work of Keskar et al.,[9] who argue that SGD gets its good generalization properties from the fact that it finds "shallow" as opposed to "sharp" minima. The basic insight is that SGD tends to jump out of minima without broad basins around them and only really settle into minima with large attractors, which tend to be the exact sort of minima that generalize. Keskar et al. use the following diagram to explain this phenomena:

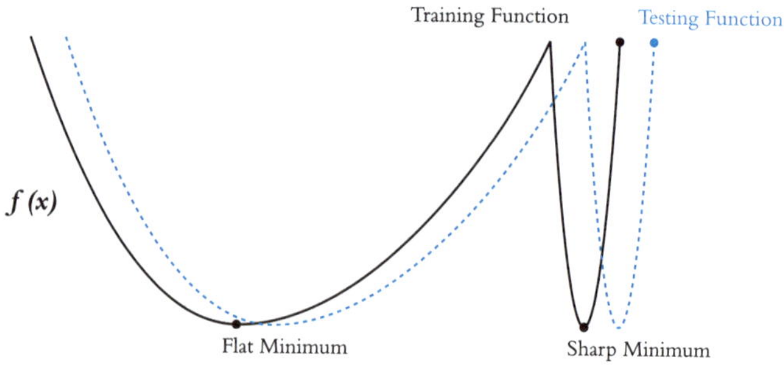

The more recent work of Dinh et al. in "Sharp Minima Can Generalize For Deep Nets",[10] however, calls the whole shallow vs. sharp minima hypothesis into question, arguing that deep networks have really weird geometry that doesn't necessarily work the way Keskar et al. want it to. (EDIT: Maybe not. See this comment[11] for an explanation of why Dinh et al. doesn't necessarily rule out the shallow vs. sharp minima hypothesis).

Another idea that might help here is Frankle and Carbin's "Lottery Ticket Hypothesis",[12] which postulates that large neural networks work well because they are likely to contain random subnetworks at initialization (what they call "winning tickets") which are already quite close to the final policy (at least in terms of being highly amenable to particularly effective training). My guess as to how double descent works if the Lottery Tickets Hypothesis is true is that in the interpolation regime SGD gets to just focus on the wining tickets and ignore the others—since it doesn't have to use the full model capacity—whereas on the interpolation threshold SGD is forced to make use of the full network (to get the full model capacity), not just the winning tickets, which hurts generalization.

That's just speculation on my part, however—we still don't really understand the inductive biases of our models, despite the fact that, as double descent shows, inductive biases are the reason that modern ML (that is, the interpolation regime) works as well as it does. Furthermore, as I noted previously, inductive biases are highly relevant to the likelihood of possible dangerous phenomenon such as mesa-optimization[4] and pseudo-alignment.[5] Thus, it seems quite important to me to do further work in this area and really understand our models' inductive biases, and I applaud Preetum et al. for their exciting work here.

(1) *Preetum Nakkiran, Gal Kaplun, Yamini Bansal, Tristan Yang, Boaz Barak, Ilya Sutskever. Deep Double Descent: Where Bigger Models And More Data Hurts.*

(2) *Preetum Nakkiran, Gal Kaplun, Yamini Bansal, Tristan Yang, Boaz Barak, Ilya Sutskever. Deep Double Descent. 5th Dec 2019.*

(3) *Mikhail Belkin, Daniel Hsu, Siyuan Ma, Soumik Mandal. Reconciling modern machine learning practice and the bias-variance trade-off. 10th Sep 2019.*

(4) *Evan Hubinger, Chris van Merwijk, vlad_m, Joar Skalse, Scott Garrabrant. Conditions for Mesa-Optimization. 1st Jun 2019.*

(5) *Evan Hubinger, Chris van Merwijk, vlad_m, Joar Skalse, Scott Garrabrant. The Inner Alignment Problem. 3rd Jun 2019.*

(6) *Paul Christiano. What does the universal prior actually look like? 30th Nov 2016.*

(7) *Evan Hubinger. Are minimal circuits deceptive? 7th Sep 2019.*

(8) *Evan Hubinger, Chris van Merwijk, vlad_m, Joar Skalse, Scott Garrabrant. Deceptive Alignment. 5th Jun 2019.*

(9) *Nitish Shirish Keskar, Dheevatsa Mudigere, Jorge Nocedal, Mikhail Smelyanskiy, Ping Tak Peter Tang. On Large-Batch Training For Deep Learning: Generalization Gap And Sharp Minima. 9th Feb 2017.*

(10) *Laurent Dinh, Razvan Pascanu, Samy Bengio, Yoshua Bengio. Sharp Minima Can Generalize For Deep Nets. 15th May 2017.*

(11) *interstice (comment). Understanding "Deep Double Descent". 5th Dec 2019.*

(12) *Jonathan Frankle, Michael Carbin. The Lottery Ticket Hypothesis: Finding Sparse, Trainable Neural Networks. 4th Mar 2019.*

How to Ignore Your Emotions (while also thinking you're awesome at emotions)

Since middle school I've generally thought that I'm pretty good at dealing with my emotions, and a handful of close friends and family have made similar comments. Now I can see that though I was great at never flipping out, I was decidedly not good at healthy emotional processing.

Relevant context

When I was a kid I adopted a strong "fix it or stop complaining about it" mentality. This applied among other things, to stress and worry. "Either address the problem you're worried about, or quit worrying about it"! Also, being a kid, I had a limited capacity to fix many of my problems, and as such I was often exercising the "stop worrying about it" option.

Another thing about me is that I was a massive book worm and loved to collect "obvious mistakes" that heroes and villains would make. My theory was, "Know all the traps, and then just don't fall for them". That, plus the sort of books I read, meant that I "knew" it was a big no-no to ignore or repress your emotions. Luckily, since I knew you shouldn't repress your emotions, I "just didn't" and have lived happily ever after!

...

...

yeah nopes.

Wiggling ears

It can be really hard to teach someone to move in a way that is completely new to them. I teach parkour, and sometimes I end up having conversations like:

Me: "Do the shock absorbing thing with your legs"!

Student: "What's the shock absorbing thing"?

Me: "... uh, you know... the thing were your legs... absorb shock"?

It's hard to know how to give cues that will lead to someone making the right mental/muscle connection. Learning new motor movements is somewhat of a process of flailing around in the dark, until some feedback mechanism tells you you did it right (a coach giving you the thumbs up, it being visually obvious in the mirror, the jump not hurting anymore, etc.). Wiggling your ears is a nice concrete version of:

a) a movement most people's bodies are capable of, and
b) one where most people feel like it's impossible.

Claim: learning mental and emotional skills has a similar "flailing around in the dark" aspect. There are the mental and emotional controls you've practiced, and those just feel like moving your arm. Natural, effortless, atomic. But there are other moves, which you are *totally capable of*, which seem impossible because you don't know how your "control panel" connects to that output. It feels like trying to wiggle your ears.

Why "ignore" and "deal with" looked the same

So, young me was upset that the grub master for our camping trip forgot half the food on the menu, and all we had for breakfast was milk. I couldn't "fix it" given that we were in the woods, so my next option was "stop feeling upset about it". So I reached around in the dark of my mind, and Oops, the "healthily process feelings" lever was *right next to* the "stop listening to my emotions" lever.

The end result? "Wow, I decided to stop feeling upset, and then I stopped feeling upset. I'm so fucking good at emotional regulation"!!!!

My current understanding of what was happening in that moment, is that I'd substituted[1] "is there a monologue of upsetness in my conscious mental loop"? for "am I feeling upset"? So from my perspective, it just felt like I was very in control of my feelings. Whenever I wanted to stop feeling something, I could. When I thought of ignoring/repressing emotions, I imagined trying to cover up something that was there, maybe with a story or lie. Or I thought if you poked around ignored emotions there would be a response of anger or annoyance. I at least expected that if I was ignoring my emotions, that if I got very calm and then asked myself, "Is there anything that you're feeling"?, that I would get an answer.

Again, the assumption was, "If it's in my mind, I should be able to notice if I look". This ignored what was actually happening, which was that I was cutting the phone lines so my emotions couldn't talk to me in the first place. Actually, the phone lines metaphor is a bit off, so here's a better one.

Parent-child model

My self-concept and conscious mind are the parent. Emotions are young children that run up to the parent to tell them something. Sometimes the child runs up to complain, "Heeeeeeeeeey I'm huuuuuuungry"! My emotional management was akin to the parenting style of slapping the child and saying, "Being hungry would suck, so you aren't hungry".

Yikes.

I know full well that you can't slap someone into having a full stomach, but you *can* slap someone into not bringing their complaints to you.

I've experienced this metaphor *directly extending* to my internal world. My emotions/ sub-agents[2] aren't stupid. They learned that telling me, "Hey, you're concerned about your relationship with your friend", "Hey, we really don't like getting laughed at", or "Hey, we're concerned that this bad thing is going to happen indefinitely" would result in getting slapped. So they learned to stay quiet.

This got to the point where I'd feel awesome and great during my busy week, and then "mysteriously" and "for no reason" feel an amorphous blob of gray badness on the weekends. I had various social and emotional needs that weren't being met, but I didn't realize that. I quite intensely tried to introspect to see if this gray blob was "about anything", but only heard quiet static. This was me being the angry parent with their kids having a dinner of half a slice of bread each, shouting, "Is anyone hungry?! Huh??! No? GREAT".

Owwwwwwwwwwwwwwwwwwwwwwwwwwwwwwwwwwwww

... and now?

When I was a kid, my desire to "not worry if it was useless" was mostly one of "people who worry seem to be in pain, I'd prefer to not be in pain". Over time, it turned into a judgmental worldview. How wasteful and useless to be embarrassed/worried/scared/etc. This was the transition from a naive parent telling their kid, "Hmmm, have you tried not being hungry"? to the angry parent shouting, "You won't be hungry in my house"!! (One might wonder how exactly that transition from naive to judgmental happened. That's a whole other story for a different post).

Over the past year I've haphazardly freestyled towards opening up emotional communication with myself, and I've made progress. I'm still not sure what "healthy emotional processing" looks like, but I've gotten HUGE gains from just being able to sit with the fact that I'm feeling something, and hug the metaphorical child that brought that emotion instead of slapping them.

I guess the biggest thing I wanted to impart with this piece was

1. the parent-child model, but also
2. that ignoring your emotions can start as a simple innocent mistake.

Related. Recently, I recall seeing a sentiment in a thread on LW that the biggest barrier to rational discourse is creating environments where everyone feels safe thinking. Extend that idea to the mind. The biggest barrier to rational thinking is organizing your mind such that it's safe to think. I still promote and admire "look towards the truth, even if it hurts", but I also now see that if you don't spend enough resources on addressing that hurt, the hurt parts of yourself can and *will* take measures to protect themselves. If you want to seek the truth, you should treat yourself well.

 Originally published 31st July 2019

(1) *Question Substitution. Hazard. jhazard.com. July 18 2019.*

(2) *Sequence introduction: non-agent and multiagent models of mind. Kaj Sotala. 7th Jan 2019.*

John S. Wentworth

Paper-Reading for Gears

Lesswrong has a fair bit of advice on how to evaluate[1] the claims made in scientific papers. Most of this advice seems to focus on a single-shot use case - e.g. a paper claims that taking hydroxyhypotheticol reduces the risk of malignant examplitis, and we want to know how much confidence to put on the claim. It's very black-box-y: there's a claim that if you put X (hydroxyhypotheticol) into the black box (a human/mouse) then Y (reduced malignant examplitis) will come out. Most of the advice I see on evaluating such claims is focused around statistics, incentives, and replication - good general-purpose epistemic tools which can be applied to black-box questions.

But for me, this black-box-y use case doesn't really reflect what I'm usually looking for when I read scientific papers.

My goal is usually not to evaluate a single black-box claim in isolation, but rather to build a gears-level[2] model of the system in question. I care about whether hydroxyhypotheticol reduces malignant examplitis only to the extent that it might tell me something about the internal workings of the system. I'm not here to get a quick win by noticing an underutilized dietary supplement; I'm here for the long game, and that means making the investment[3] to understand the system.

With that in mind, this post contains a handful of thoughts on building gears-level models from papers. Of course, general-purpose epistemic tools (statistics, incentives, etc) are still relevant - a study which is simply wrong is unlikely to be much use for anything. So the thoughts and advice below all assume general-purpose epistemic hygiene as a baseline - they are things which seem more/less important when building gears-level models, *relative to* their importance for black-box claims.

I'm also curious to hear other peoples' thoughts/advice on paper reading specifically to build gears-level models.

Get Away From the Goal

Ultimately, we want a magic bullet to cure examplitis. But the closer a paper is to that goal, the stronger publication bias and other memetic distortions will be. A flashy, exciting result picked up by journalists will get a lot more eyeballs than a failed replication attempt.

But what about a study examining the details of the interaction between FOXO, SIRT6, and WNT-family signalling molecules? That paper will not ever make the news circuit - laypeople have no idea what those molecules are or why they're interesting. There isn't really a "negative result" in that kind of study -

there's just an open question: "do these things interact, and how"? Any result is interesting and likely to be published, even though you won't hear about it on CNN.

In general, as we move more toward boring internal gear details that the outside world doesn't really care about, we don't need to worry as much about incentives - or at least not the same *kinds* of incentives.

Zombie Theories

Few people want to start a fight with others in their field, even when those others are wrong. There is little incentive to falsify the theory of somebody who may review your future papers or show up to your talk at a conference. It's much easier to say "examplitis is a complex multifactorial disease and all these different lines of research are valuable and important, kumbayah".

The result is zombie theories: theories which are pretty obviously false if you spend an hour looking at the available evidence, but which are still repeated in background sections and review articles.

One particularly egregious example I've seen is the idea that a shift in the collagen:elastin ratio is (at least partially) responsible for the increased stiffness of blood vessels in old age. You can find this theory in review[4] articles[5] and even textbooks.[6] It's a nice theory: new elastin is not produced[7] in adult vasculature, and collagen is much stiffer, so over time we'd expect the elastin to break down and collagen to bear more stress, increasing overall stiffness. But if we go look for studies which directly measure the collagen: elastin ratio in the blood vessels... we mostly find no significant change with age (rat,[8] human,[9] rat);[10] one study even[11] finds more elastin relative to collagen in older humans.

Ignore the Labels on the Box

Scientists say lots of things which are misleading, easily confused, or aren't actually supported by their experiments . That doesn't mean the experiment is useless, it just means we should ignore the mouth-motions and look at what the experiment and results actually were. As an added bonus, this also helps prevent misinterpreting what the paper authors meant.

An example: many authors assert that both (1) atherosclerosis is a universal marker of old age among humans and most other mammals, and (2) atherosclerosis is practically absent among most third-world populations. What are we to make of this? Ignore the mouth motions, look for data.[12] In this case, it looks like atherosclerosis does universally grow very rapidly with age in all populations examined, but still has much lower overall levels among third-world populations after controlling for age - e.g. ~⅓ as prevalent in most age brackets in 1950's India compared to Boston.

Read Many Dissimilar Papers: Breadth > Depth

For replication, you want papers which are as similar as possible, and establishing very high statistical significance matters. For gears-level models, you want papers which do very different things, but impinge on the same gears. You want to test a whole model rather than a particular claim, so finding qualitatively different tests is more important than establishing very high statistical significance. (You still need enough statistics to make sure any particular result isn't just noise, but high confidence will ultimately be established by incrementally updating on many different kinds of studies).

For example, suppose I'm interested in the role of thymic involution[13] as a cause of cancer. The thymus is an organ which teaches new adaptive immune cells (T-cells) to distinguish our own bodies from invaders, and it shrinks ("involutes") as we age.

Rather than just looking for thymus-cancer studies directly, I move away from the goal and look for general information on the gears of thymic involution. Eventually I find that castration of aged mice[14] (18-24 mo) leads to complete restoration of the thymus in about 2 weeks. The entire organ completely regrows, and the T-cells return to the parameters seen in young mice. (Replicated here[15]) Obvious next question: does castration reduce cancer? It's used as a treatment for e.g. prostate cancer, but that's (supposedly) a different mechanism. Looking for more general results turns up this century-old study,[16] which finds that castration prevents age-related cancer in mice - and quite dramatically so. Castrated old mice' rate of resistance to an implanted tumor was ~50%, vs ~5% for controls. Even more interesting: castration did not change the rate of tumor resistance in young mice - exactly what the thymus-mediation theory would predict.

This should *not*, by itself, lead to very high confidence about the castration => thymus => T-cell => cancer model. We need more qualitatively different studies (especially in humans), and we need at least a couple studies looking directly at the thymus => cancer link. But if we find a bunch of different results, each with about this level of support for the theory, covering interventions on each of the relevant variables, then we should have reasonable confidence in the model. It's not about finding a single paper which proves the theory for all time; it's about building up Bayesian evidence from many qualitatively different studies.

Mediation is Everything

Everything is correlated with everything else;[17] any intervention changes everything.

That said, very few things are *directly* connected; the main value is finding variables which *mediate* causal influence. For instance, maybe hydroxyhypotheticol usually reduces malignant examplitis, but most of the effect goes away if we hold hypometabolicol levels constant. That's a powerful finding: it establishes that hypometabolicol is one of the internal gears between hydroxyhypotheticol and examplitis.

If I had to pick the single most important guideline for building gears-level models from papers, this would be it: mediation is the main thing we're looking for.

 Originally published 4th December 2019

(1) *Eli Tyre. How do you assess the quality / reliability of a scientific study? 2019.*

(2) *John S. Wentworth. Gears vs Behavior. 19th September 2019.*

(3) *John S. Wentworth. Gears-Level Models are Capital Investments. 22nd November 2019.*

(4) *Jason C. Kovacic, MD, PhD; Pedro Moreno, MD; Elizabeth G. Nabel, MD; Vladimir Hachinski, MD, DSc; Valentin Fuster, MD, PhD. Cellular Senescence, Vascular Disease, and Aging. Part 1. 2011.*

(5) *Jason C. Kovacic, MD, PhD; Pedro Moreno, MD; Elizabeth G. Nabel, MD; Vladimir Hachinski, MD, DSc; Valentin Fuster, MD, PhD. Cellular Senescence, Vascular Disease, and Aging. Part 2. 2011.*

(6) *Paola S. Timiras. Physiological Basis of Aging and Geriatrics, 3rd Edition. 2002.*

(7) *Janet T. Powell, Nicholas Vine, Margot Crossman. One the accumulation of D-aspartate in elastin and other proteins of the ageing aorta. 1992.*

(8) *Kung-Ying Tang Kao, Thomas H. McGavack. Connective Tissue: I. Age and Sex Influence on Protein Composition of Rat Tissues. 1959.*

(9) *Yasuhiro Hosoda, Koichi Kawano, Fumihiro Yamasawa, et al. Age-Dependent Changes of Collagen and Elastin Content in Human Aorta and Pulmonary Artery. 1984.*

(10) *Annemarie Brüel, Hans Oxlund. Changes in biomechanical properties, composition of collagen and elastin, and advanced glycation endproducts of the rat aorta in relation to age. Atherosclerosis, Volume 127, Issue 2. 1996.*

(11) *M. Anne Cattell, John C. Anderson, Philip S. Hasleton. Age-related changes in amounts and concentrations of collagen and elastin in normotensive human thoracic aorta. Clinica Chimica Acta, Volume 245, Issue 1. 1996.*

(12) *K.S. Mathur, M.D., F.R.C.P., N. 1I. Patney, M.D., and V. Kumar, M.D. July 1961.*

(13) *Wikipedia.org. Thymic involution.*

(14) *Jayne S. Sutherland, Gabrielle L. Goldberg, Maree V. Hammett, Adam P. Uldrich, Stuart P. Berzins, Tracy S. Heng, Bruce R. Blazar, Jeremy L. Millar, Mark A. Malin, Ann P. Chidgey and Richard L. Boyd. Activation of Thymic Regeneration in Mice and Humans following Androgen Blockade. August 15, 2005.*

(15) *Daniel H. D. Gray, Natalie Seach, Tomoo Ueno, Morag K. Milton, Adrian Liston, Andrew M. Lew, Christopher C. Goodnow, Richard L. Boyd. Developmental kinetics, turnover, and stimulatory capacity of thymic epithelial cells. December 1, 2006.*

(16) *James B. Murphy, Ernest Sturm. Effect of Prepuberty Castration on Subsequent Cancer Implantation. Ausgust 01, 1925.*

(17) *Gwern.net/Causality. Why Correlation Usually ≠ Causation. 2014.*

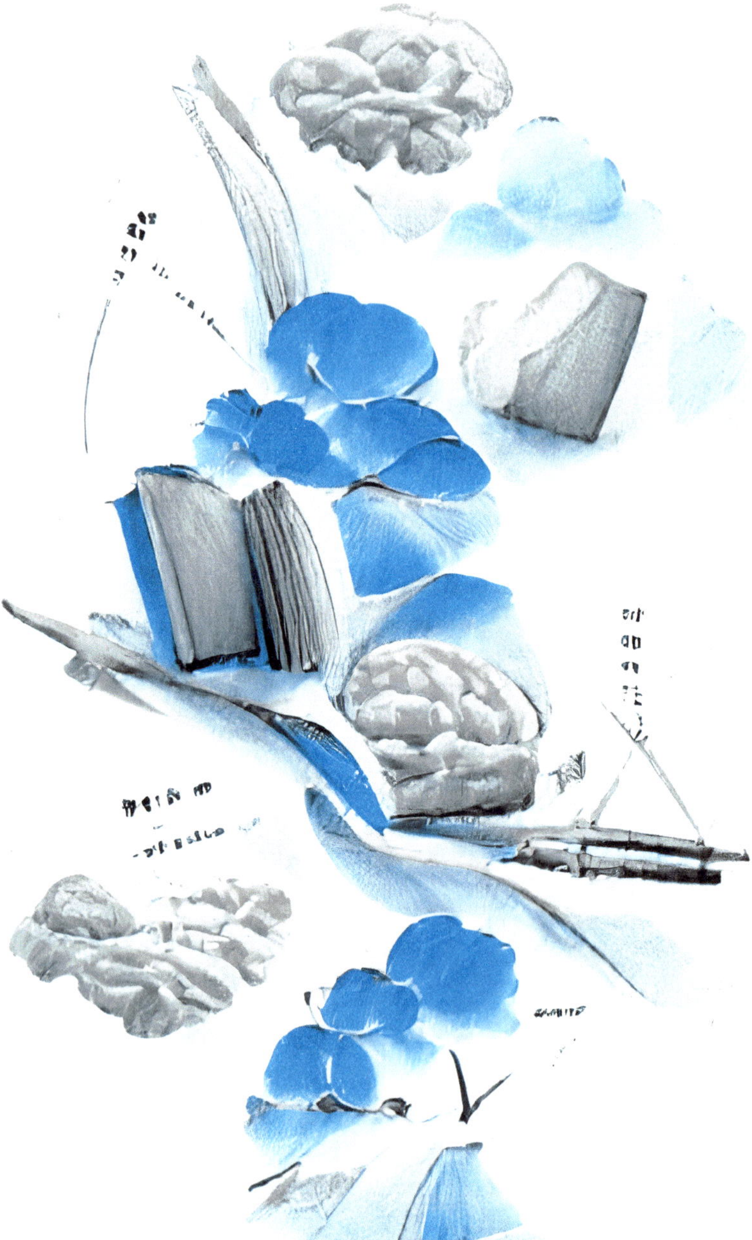

Book summary: Unlocking the Emotional Brain

If the thesis in *Unlocking the Emotional Brain*[1] (UtEB) is even half-right, it may be one of the most important books that I have read. Written by the psychotherapists Bruce Ecker, Robin Ticic and Laurel Hulley, it claims to offer a neuroscience-grounded, comprehensive model of how effective therapy works. In so doing, it also happens to formulate its theory in terms of belief updating, helping explain how the brain models the world and what kinds of techniques allow us to actually change our minds.[2] Furthermore, if UtEB is correct, it also explains why rationalist techniques such as Internal Double Crux work.[3] [4] [5]

UtEB's premise is that much if not most of our behavior is driven by emotional learning. Intense emotions generate unconscious predictive models of how the world functions and what caused those emotions to occur. The brain then uses those models to guide our future behavior. Emotional issues and seemingly irrational behaviors are generated from implicit world-models (schemas) which have been formed in response to various external challenges. Each schema contains memories relating to times when the challenge has been encountered and mental structures describing both the problem and a solution to it.

According to the authors, the key for updating such schemas involves a process of memory reconsolidation, originally identified in neuroscience. The emotional brain's learnings are usually locked and not modifiable. However, once an emotional schema is activated, it is possible to simultaneously bring into awareness knowledge contradicting the active schema. When this happens, the information contained in the schema can be overwritten by the new knowledge.

While I am not convinced that the authors are *entirely* right, many of the book's claims definitely feel like they are pointing in the right direction. I will discuss some of my caveats and reservations after summarizing some of the book's claims in general. I also consider its model in the light of an issue of a psychology/cognitive science journal devoted to discussing a very similar hypothesis.

Emotional learning

In UtEB's model, emotional learning forms the foundation of much of our behavior. It sets our basic understanding about what situations are safe or unsafe, desirable or undesirable. The authors do not quite say it explicitly, but the general feeling I get is that the subcortical emotional processes set many of the priorities for what we want to achieve, with higher cognitive functions then trying to figure out *how* to achieve it - often remaining unaware of what exactly they are doing.

UtEB's first detailed example of an emotional schema comes from the case study of a man in his thirties they call Richard. He had been consistently successful and admired at work, but still suffered from serious self-doubt and low confidence at his job. On occasions such as daily technical meetings, when he considered saying something, he experienced thoughts including "Who am I to think I know what's right"?, "This could be wrong" and "Watch out - don't go out on a limb". These prevented him from expressing any opinions.

From the point of view of the authors, these thoughts have a definite cause - Richard has "emotional learnings according to which it is adaptively *necessary* to go into negative thoughts and feelings towards [himself]". The self-doubts are a *strategy* which his emotional brain has generated for solving some particular problem.

Richard's therapist guided Richard to imagine what it would feel like if he was at one of his work meetings, made useful comments, and felt confident in his knowledge while doing so. This was intended to elicit information about what Richard's emotional brain predicted would happen if it failed to maintain the strategy of self-doubt. The book includes the following transcript of what happened after Richard started imagining the scene as instructed:

Richard: Now I'm feeling really uncomfortable, but-it's in a different way.

Therapist: OK, let yourself feel it - this different discomfort. *[Pause]*. See if any words come along with this uncomfortable feeling.

Richard: *[Pause]*. Now they hate me.

Therapist: "Now they hate me". Good. Keep going: See if this really uncomfortable feeling can also tell you why they hate you now.

Richard: *[Pause]*. Hnh. Wow. It's because… now I'm… an arrogant asshole… like my father… a totally self-centered, totally insensitive know-it-all.

Therapist: Do you mean that having a feeling of confidence as you speak turns you into an arrogant asshole, like Dad?

Richard: Yeah, exactly. Wow.

Therapist: And how do you *feel* about being like him in this way?

Richard: It's horrible! It's what I've always vowed *not* to be!

Richard had experienced his father as being assertive as well as obnoxious and hated. His emotional brain had identified this as a failure mode to be avoided: if you are assertive, then you are obnoxious and will be hated. The solution was to generate feelings of doubt so as to stop him from being too confident. This caused him suffering, but the prediction of his emotional brain was that acting otherwise would produce even worse suffering, as being hated would be a terrible fate.

UtEB describes Richard as having had the following kind of unconscious schema:

Perceptual, emotional and somatic memory of original experiences: his suffering from his father's heavily dominating, hyper-confident self-expression, plus related suffering from unmet needs for fatherly expressions of love, acceptance, understanding, validation. (This is the "raw data"; matching features in current situations are triggers of the whole schema).

A mental model or set of linked, learned constructs operating as living knowledge of a problem and a solution:

The problem: knowledge of a vulnerability to a specific suffering. Confident assertiveness in any degree inflicts crushing oppression on others and is hated by them. I would be horrible like Dad and hated by others, as he is, if I asserted my own knowledge or wishes confidently. (This is a model of how the world is, and current situations that appear relevant to this model are triggers of the whole schema).

The solution: knowledge of an urgent broad strategy and concrete tactic(s) for avoiding that suffering. Never express any confident assertiveness, to avoid being horrible and hated (general strategy and pro-symptom purpose), by vigilantly noticing any definite knowledge or opinions forming in myself and blocking them from expression by generating potently self-doubting, self-invalidating thoughts (concrete tactic and manifested symptom).

Emotional schemas can be brought to light during a variety of ways, including Focusing,[6] IFS,[7] and imagining yourself doing something and seeing what you expect to happen as a result.

But suppose that you do manage to bring up a schema which seems wrong to you. What do you do then?

Memory reconsolidation: updating the emotional learning

The formation of memory traces involves *consolidation*, when the memory is first laid out in the brain; *deconsolidation*, when an established memory is "opened" and becomes available for changes; and *reconsolidation*, when a deconsolidated memory (along with possible changes) is stored and becomes frozen again. The term "reconsolidation" is also used to refer to the general process from deconsolidation to reconsolidation; UtEB generally applies the term to mean the entire process. Unless the context indicates otherwise, I do the same.

UtEB reviews some of the history of memory research. Until 1997, neuroscientists believed that past emotional learning became permanently locked in the brain, so that memories could only consolidate, never de- or reconsolidate. More recent research has indicated that once a memory becomes activated, it is temporarily unlocked, allowing it to be changed or erased.

Starting from 2004, new studies suggested that activation alone is not sufficient to deconsolidate the memory. The memories are used to predict that things will occur in a similar fashion as they did previously. Besides just activation, there has to be a significant *mismatch* between what one experiences and what the memory suggests is about to happen. The violation of expectation can be qualitative (the predicted outcome not occurring at all) or quantitative (the magnitude of the outcome not being fully predicted). In either case, it is this prediction error which triggers the deconsolidation and subsequent reconsolidation.

The memory erasure seems to be specific to the interpretation from which the prediction was produced. For example, someone who has had an experience of being disliked may later experience being liked. This may erase the emotional generalization "I am inherently dislikable", but it will not erase the memory of the person also having been disliked.

Applied reconsolidation: an example of the schema update process

So, assuming that the model outlined above is correct, how does one apply it in practice?

From what we have discussed so far, the essential steps of erasing a learned belief (including an emotional schema) involves identifying it, activating it, and then finding a mismatch between its prediction and reality.

The first difficulty is that the beliefs involved with the schema are not necessarily consciously available at first. Richard knew that he suffered from a lack of self-esteem, but he was not aware of its reason. The process started from him describing in concrete details how this manifested: as skeptical self-talk during a daily meeting.

As he was guided to imagine what would happen if he didn't have those thoughts and acted confidently, his therapist was seeking to retrieve the implicit schema and bring it into consciousness so that its contents would become available for access. Once it had been retrieved, the therapist and Richard worked together to express the belief in the schema in maximally emotional language:

> "Feeling *any* confidence means I'm arrogant, self-centered, and totally insensitive like Dad, and people will hate me for it, so I've got to *never* feel confident, ever".

The authors have developed a therapeutic approach called Coherence Therapy, whose steps closely follow the steps of the memory reconsolidation process. The example of Richard is from this school of therapy.

In Coherence Therapy (as well as related approaches, such as Internal Family Systems),[7] one initially avoids any impulse to argue with or disprove the retrieved schema. This would risk it being pushed away before it has become sufficiently activated to allow for reconsolidation.

Instead, one stays with it. Richard was given a card with the above phrase and instructed to review it every day until the next therapy session, just feeling the extent to which it felt true to him. This served to further integrate access to the schema in question, making it better consciously available.

Two weeks later, Richard had frequently noticed his self-doubt, used it as a prompt for reading the card, and experienced its description ringing true as a reason for his thoughts. When speaking with his therapist, he mentioned a particular event which had stuck in his memory. In a recent meeting, he had thought of a solution to a particular problem, but then kept quiet about it. A moment later, another person had spoken up and suggested the same solution in a confident manner. Looking around, Richard had seen the person's solution and confidence being received positively by the others. Richard had been struck by how that reaction differed from what his schema predicted would happen if he had made the same suggestion in that tone.

Because Richard had made the implicit assumptions in his schema explicit, he was able to consciously notice a situation which seemed to violate those assumptions: a prediction mismatch. His therapist recognized this as a piece of contradictory knowledge which could be used to update the old schema. The therapist then guided Richard through a process intended to activate the old schema while bringing the contradictory information into awareness, triggering a reconsolidation process.

The therapist first instructed Richard to mentally bring himself back to the situation where he had just thought of the solution, but held it back. To properly activate the schema, the therapist guided Richard's attention to the purpose behind his reluctance and Richard's certainty of any confidence making him disliked. Next, the therapist told Richard to re-live what happened next: the other person making the same suggestion, and the other people in the room looking pleased rather than angry.

The book then has a transcript of the therapist guiding Richard to repeat this juxtaposition of the old schema and the disconfirming experience (italicized brackets in the original):

> **Therapist:** Stay with that. Stay with being surprised at what you're seeing—surprised because in your life, you've had such a definite knowing that saying something confidently to people will always come across like Dad, like an obnoxious know-it-all, and people will hate that. That's what you know, yet at the same time, here you're seeing that saying something confidently isn't always like Dad, and then people are fine with it. And it's quite a surprise to know that. *[That was an explicit prompting of another side-by-side experiencing of the two incompatible knowings, with the therapist expressing empathy for both, with no indication of any favoring of one knowing over the other. The therapist paused for several seconds, then asked]:*

Does it feel true to describe it like that? Your old knowing right alongside this other new knowing that's so different?

Richard: *[Quietly, seeming absorbed in the experience].* Yeah.

Therapist: *[Softly].* All along, it seemed to you that saying something confidently could be done only in Dad's dominating way of doing it, and now suddenly you're seeing that saying something confidently can be done very differently, and it feels fine to people. *[This was another deliberate repetition of the same juxtaposition experience].*

Richard: Yeah.

Therapist: Mm-hm. *[Silence for about 20 seconds].* So, how is it for you be in touch with both of these knowings, the old one telling you that anything said with confidence means being like Dad, and the new one that knows you can be confident in a way that feels okay to people? *[Asking this question repeated the juxtaposition experience yet again, and, in addition, the "how is it" portion of the question prompted Richard to view the experience with mindful or metacognitive awareness, while remaining in the experience].*

Richard: It's sort of weird. It's like there's this part of the world that I didn't notice before, even though it's been right there.

Therapist: I'm intrigued by how you put that. Sounds like a significant shift for you.

Richard: Yeah, it is. Huh.

Therapist: You're seeing both now, the old part of the world and this other part of the world that's new, even though it was right there all along. *[That cued the juxtaposition experience for a* *fourth time, followed by silence for about 30 seconds].* So, keep seeing both, the old part and the new part, when you open your eyes in a few seconds and come back into the room with me. *[Richard soon opens his eyes and blinks a few times].* Can you keep seeing both?

Richard: Yeah.

Therapist: What's it like to see both and feel both now? *[With the transformation sequence complete, this question begins the next step of verification— Step V—because it probes for whether the target learning still exists as an emotional experience].*

Richard: *[Pause, then sudden, gleeful laughter].* It's kind of funny! Like, what? How could I think that? *[This is an initial marker indicating that the pro-symptom schema may have been successfully disconfirmed, depotentiated, and dissolved by the transformation sequence].*

Therapist: Do you mean, how could you think that simply saying what you know, or mentioning some good idea that you've had, would make you seem arrogant, insensitive and dominating like Dad and be hated for it?

Richard: *[Laughing again].* Yeah!

Afterwards, the therapist and Richard wrote a new card together, which Richard was told to review daily:

All along it's been so clear that if I confidently say what I know, I will always come across as arrogant, insensitive, and dominating like Dad, and be hated for it. And it's so weird, looking around the room and seeing that it doesn't come across like that.

The purpose of the card was to provide additional juxtaposition experiences between the old schema and the new knowledge. While the original transformation sequence might have been enough to eliminate the old schema, the schema might also have been stored in the context of many different situations and contained in several memory systems. In such a situation, further juxtapositions would have helped deal with it.

In a follow-up meeting, Richard reported having lost the feelings of self-doubt, and that speaking up no longer felt like it was any big deal. To verify that the old schema really had lost its power, the therapist tried deliberately provoking his old fears again:

> Dropping his voice to a quieter tone, the therapist added, "But tell me, when you have something to say and just say it, what about the danger of coming across as a know-it-all, like Dad, and being hated for that? What about your fear of that and how urgent it is to protect yourself from that"? [...]
>
> Richard took in the question, gazed at the therapist in silence for a few seconds, and then replied, "Well, I don't know what to tell you. All I can say is, that doesn't trouble me any more. And hearing you say it, it seems a little strange that it ever did—like, what was my problem"?

Applied reconsolidation: the schema update process in general

Now that we have looked at a specific example, we can look at a more general version of the process.

Accessing sequence

In Coherence Therapy, the *accessing sequence* is the preliminary phase of making both a person's implicit schema and some disconfirming knowledge accessible, so that they can be used in the juxtaposition process:

1. *Symptom identification.* Establishing which specific symptoms the person regards as problematic, and when and where they manifest. In Richard's case, the general symptom was a lack of confidence, which specifically manifested as negative self-talk in meetings.
2. *Retrieval of target learning.* Bringing into explicit awareness the purpose behind the symptoms. This can then be used to guide the search for disconfirming knowledge, as well as accessing the original schema in order to reconsolidate it. In Richard's case, the purpose was to avoid expressing confidence in a way that would make people hate him.
3. *Identification of disconfirming knowledge.* Identifying some past or present experience which directly contradicts the original learning. This knowledge does not necessarily need to feel "better" or "more positive" than the old one, just as long as it is mutually exclusive with the old one. In Richard's case, the disconfirming knowledge was the experience of his co-worker confidently proposing a solution and being well-received.

Erasure sequence

Once both the target schema and the disconfirming knowledge are known, the erasure steps can be applied to update the learning:

1. *Reactivation of the target schema.* Tapping into the felt truth of the original learning, experiencing it as vividly as possible.
2. *Activation of disconfirming knowledge, mismatching the target schema.* Activating, at the same time, the contradictory belief and having the experience of simultaneously believing in two different things which cannot both be true.
3. *Repetitions of the target-disconfirmation pairing.*

Something that the authors emphasize is that when the target schema is activated, there should be no attempt to explicitly argue against it or disprove it, as this risks pushing it down. Rather, the belief update happens when one experiences their old schema as vividly true, while *also* experiencing an entirely opposite belief as vividly true. It is the juxtaposition of believing X and not-X at the same time, which triggers an inbuilt contradiction-detection mechanism in the brain and forces a restructuring of one's belief system to eliminate the inconsistency.

The book notes that this distinguishes Coherence Therapy from approaches such as Cognitive Behavioral Therapy, which is premised on treating some beliefs as intrinsically irrational and then seeking to disprove them. While UtEB does not go further into the comparison, I note that this is a common complaint that I have heard of CBT: that by defaulting to negative emotions being caused by belief distortions, CBT risks belittling those negative emotions which are actually produced by *correct* evaluations of the world.

I would personally add that not only does treating all of your beliefs - including emotional ones - as provisionally valid seem to be a requirement[8] for actually updating them, this approach is also good rationality practice. After all, you can only seek evidence to test a theory, not confirm it.[9]

If you notice different parts of your mind having conflicting models of how the world works, the correct epistemic stance should be that you are trying to figure out which one is true - not privileging one of them as "more rational" and trying to disprove the other. Otherwise it will be unavoidable that your preconception will cause you to dismiss as false beliefs which are actually true. (Of course, you can still reasonably *anticipate* the belief update going a particular way - but you need to take seriously at least the *possibility* that you will be shown wrong).

This can actually be a relief. Trying to stack the deck towards receiving favorable evidence would just also sabotage the brain's belief update process. So you might as well give up trying to do so, relax, and just let the evidence come in.[9]

I speculate that this limitation might also be in place in part to help avoid the error where you decide which one of two models is more correct, and then discard the other model entirely. Simultaneously running two contradictory schemas at the same time allows both of them to be properly evaluated and merged[10] rather than one of them being thrown away outright. I suspect that in Richard's case, the resulting process didn't cause him to entirely discard the notion that some behaviors will make him hated like his dad was – it just removed the overgeneralization which had been produced by having too little training data as the basis of the schema.[11]

Of course, this means that there does need to be some contradictory information available which *could* be used to disprove the original schema. One might have a schema for which no disconfirmation is available because it is correct, or a schema which might or might not be correct but which is making things worse and cannot easily be disconfirmed. UtEB mentions the example of a man, "Tőmas", who had a desire to be understood and validated by someone important in his life. Tómas remarked that a professional therapist who was being paid for his empathy could never fulfill that role. The update contradicting the schema that nobody in his life really understood him, would have to come from someone actually in his life.

Another issue that may pop up with the erasure sequence is that there is another schema which predicts that, for whatever reason, running this transformation may produce adverse effects.[12] In that case, one needs to address the objecting schema first, essentially be carrying out the entire process on it before returning to the original steps. (This is similar to the phenomenon in e.g. Internal Family Systems[7], where objecting parts may show up and have their concerns addressed before work on the original part can proceed).

Verification step

Finally, after the erasure sequence has been run, one seeks to verify that lasting change has indeed happened and that the target schema has been transformed. UtEB offers the following behavioral markers as signs that a learning which has previously generated emotional responses has in fact been erased:

- "A specific emotional reaction abruptly can no longer by re-activated by cues and triggers that formerly did so or by other stressful situations".
- "Symptoms of behavior, emotion, somatics, or thought that were expressions of that emotional reaction also disappear permanently".

- "Non-recurrence of the emotional reaction and symptoms continues effortlessly and without counteractive or preventive measures of any kind".

The authors interpret current neuroscience to say that only memory reconsolidation can produce these kinds of markers. They cannot be produced by counteractive or competitive processes, such as trying to learn an opposite habit to replace a neurotic behavior. Counteractive processes are generally fragile and susceptible to relapse. When these markers are observed in clinical work, UtEB argues that one may infer that reconsolidation has led to the original learning being replaced.

Further examples

For additional examples of the schema update process, I recommend reading the book, which contains several more case studies of issues which were dealt with using this approach. Here's a brief summary of the most detailed ones (note that some of these examples are actually more detailed and include additional complications, such as more than one symptom-producing schema; I have only summarized the most prominent ones to give a taste of them):

- "Charlotte". *Issue*: obsessive attachment to a former lover. *Schema*: "It would be much better if I was merged with my lover". *Contradictory knowledge*: The harm caused by not having boundaries.
- "Ted". *Issue*: an inability to hold a steady job and a general lack of success in life. *Schema*: "If my life is a mess, my father will be forced to admit how badly he screwed up as a parent". *Contradictory knowledge*: Realizing that Ted's father would never admit failure, no matter what.
- "Brenda". *Issue*: stage fright when having a leading role in an upcoming play. *Schema*: being on the stage in front of an

audience means being unable to get off, causing helplessness similar to when Brenda was in a car with her alcoholic father and couldn't get out. *Contradictory knowledge*: re-imagining the scene and the way how Brenda *could* actually have gotten out of the car.

- "Travis". *Issue*: inability to experience intimate emotional closeness in relationships. *Schema*: "Nobody will pay attention to how I feel or give me understanding for how I'm hurting. I don't matter, and I'm all on my own". *Contradictory knowledge*: the therapist's empathic presence and listening.
- "Regina". *Issue*: strong anxiety and panic during/after interacting with other people. *Schema*: "I'm acceptable and lovable only if I do everything *perfectly*". *Contradictory knowledge*: Regina's Uncle Theo loves her regardless of what imperfections she might have.
- "Carol". *Issue*: wanting to avoid sex with her husband despite feeling emotionally close to him. *Schema*: engaging in any sexuality means being overtly sexual and harming Carol's daughter, in the way that Carol was harmed by her mother's overt sexuality. *Contradictory knowledge*: once the schema was made conscious, it activated the brain's spontaneous mismatch detection mechanisms and started to feel silly.

As the last item suggests, sometimes just making a schema explicit is enough to start to dismantle it. The authors suggest that the brain has a built-in detection system which compares any consciously experienced beliefs for inconsistencies with other things that a person knows, and can spontaneously create juxtaposition experiences by bringing up such inconsistent information. They suggest that therapies which are based on digging up previously unconscious material, but which do not have an explicit juxtaposition step, work to the extent that the uncovered material happens to trigger this spontaneous mismatch detection. (We already saw this happening with Richard - once his underlying schema had been made conscious, he was startled to later notice what seemed like a contradiction).

One may note the connection to the model in *Consciousness and the Brain*[13] that when some subsystem in the brain manages to elevate a mental object into the content of consciousness, multiple subsystems will synchronize their processing around that object. If the object is an explicit belief, then any subsystem which is paying attention to that object may presumably detect inconsistencies with that subsystem's own models.

Besides these case studies from Coherence Therapy, the authors also analyze published case studies from Accelerated Experiental Dynamic Psychotherapy, Emotion-Focused Therapy, Eye-Movement Desensitization and Reprocessing, and Interpersonal Neurobiology. They try to show how these cases also carried out a juxtaposition process, even if the theoretical frameworks of those therapies did not explicitly realize it. It is the claim of the authors that any therapy which causes lasting emotional change does it through reconsolidation. Finally, the book contains four essays from other therapists (using Coherence Therapy and EMDR), who analyze some of their own case studies.

Evaluating the book's plausibility

Now that we have looked at the book's claims, let's look at whether we should believe in them.

It is unclear to me how reliable the neuroscience results are; the authors cite a number of studies, but each individual claim only references a relatively small number of them.

On a brief look, I could not find any reviews or papers that would have directly made a critical assessment of the book's model. However, I found something that might be even better.

Behavioral and Brain Sciences is a respected journal covering subject areas across the cognitive sciences. BBS publishes "target articles" which present some kind of a thesis or review about a particular topic, together with tens of brief commentaries which respond to the target article, and a final response by the target article's authors to the commentaries.

In 2015, four prestigious (with a total of 500 published research articles between them) psychologists published a BBS target article, *Memory reconsolidation, emotional arousal, and the process of change in psychotherapy: New insights from brain science*[14] (Lane et al. 2015). While the exact model that they outline has a number of differences from the UtEB model, the core idea is the same: that therapeutic change from a wide variety of therapeutic approaches, "including behavioral therapy, cognitive-behavioral therapy, emotion-focused therapy, and psychodynamic psychotherapy, results from the updating of prior emotional memories through a process of reconsolidation that incorporates new emotional experiences".

One interesting difference was that Lane et al. describe emotional schemas somewhat differently. In their model, the schemas form memory structures with three mutually integrated components: emotional responses, episodic/autobiographical memories, and semantic structures (e.g. abstract beliefs which generalize over the various incidents, such as the claim that "people are untrustworthy"). Any of these components can be used as an entry point to the memory structure, and can potentially update the other components through reconsolidation. They hypothesize that different forms of therapy work by accessing different types of components: e.g. behavior therapy and emotion-focused therapy access emotional responses, conventional psychoanalysis uses access to biographical memories, and cognitive behavioral therapy accesses semantic structures.

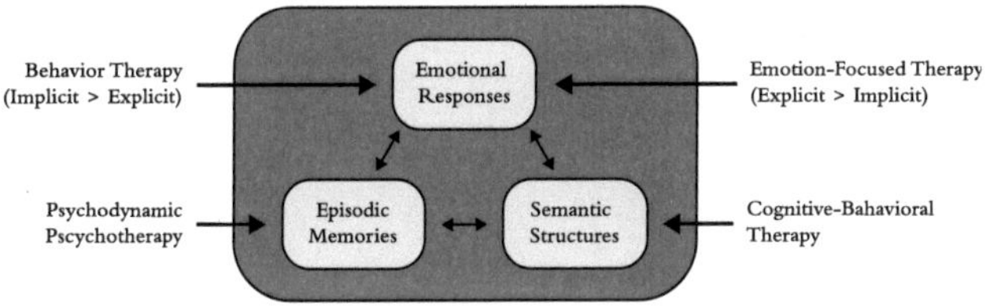

I read the target article, all the commentaries, and the responses. Given the similarities between Lane et al.'s model and the UtEB model, I think we can consider the responses to Lane et al. to generally offer a useful evaluation of the UtEB model as well.

One significant difference which needs to be noted is that Lane et al.'s model of memory reconsolidation does not mention the requirement for a prediction mismatch before reconsolidation can happen. This was remarked on in the response from UtEB's authors. In their counter-response, Lane et al. noted UtEB's model to be highly compatible with theirs, and remarked that further research is needed to nail down the conditions which make reconsolidation the most effective.

The other responses to Lane et al. were mostly from psychologists, psychiatrists, and neuroscientists, but also included the occasional economist, philosopher, philologist and folklorist. Several of the responses were generally positive and mostly wanted to contribute additional details or point out future research directions.

However, there were also a number of skeptical responses. A common theme which emerged from several concerned the limitations of the current neuroscience research on memory reconsolidation. In particular, most of the studies so far have been carried out on rats, and specifically testing the elimination of a fear response to electric shocks. As one of the responses points out, "neither the stimuli nor the subjects are generalizable to the kind of rich autobiographical memories involved in therapy." A number also raised the question whether all therapeutic change really involves reconsolidation, as opposed to some related mechanism, such as creating new memory structures which compete with the original as opposed to replacing it.

My non-expert reading is that the critical responses are right in that a gap remains between the clinical and behavioral findings on the one hand, and the neuroscience findings on the other. There are various patterns which can be derived from psychological research and clinical therapy experience, and a small number of neuroscience findings which establish the existence of something that *could* explain those patterns. However, the neuroscience findings have only been established in a rather narrow and limited context; the connection between them and the higher-level patterns is a plausible link, but it remains speculative nonetheless.

Personally I consider the book's model tentatively promising, because it seems to explain many observations which I had independently arrived at *before* reading it. For example, I had noticed an interesting thing with anxieties, where I let e.g. a sensation of social anxiety stay active in my mind, neither accepting it as truth nor pushing it away while I went to do social things. This would then cause the anxiety to update,[15] making me feel less anxious if it was indeed the case that the social interaction was harmless. This fits nicely together with the framework of an activated memory structure becoming open to reconsolidation and then being updated by a prediction mismatch (the situation not being as bad as expected).

Likewise, in my post *Integrating disagreeing subagents*,[12] I reviewed a variety of rationality and therapeutic techniques, and suggested that they mostly worked either by merging or combining two existing models that a person's brain already had, or augmenting the existing models by collecting additional information.

In particular, I considered an example from cognitive behavioral therapy, where a man named Walter was feeling like he was impossible to be in a relationship with after he had broken up with his boyfriend. At the same time, he did not think that someone *else* breaking up with their partner was an indication of them being impossible to be in a relationship with. He and his therapist role-played an interaction where the therapist pretended to be a friend who had recently broken up, and Walter explained why this did not make the friend a relationship failure. In the process of doing so, Walter suddenly realized that he wasn't a failure, either.

I commented:

> Walter was asked whether he'd say something harsh to a friend, and he said no, but that alone wasn't enough to improve his condition. What did help was putting him in a position where he had to really think through the arguments for why this is irrational in order to convince his friend, and then, after having formulated the arguments once himself, get convinced by them himself.

> In terms of our framework, we might say that a part of Walter's mind contained a model which output a harsh judgment of himself, while another part contained a model which would output a much less harsher judgment of someone else who was in otherwise identical circumstances. Just bringing up the existence of this contradiction wasn't enough to change it: it caused the contradiction to be noticed, but didn't activate the relevant models extensively enough for their contents to be reprocessed.
>
> But when Walter had to role-play a situation where he thought of himself as actually talking with a depressed friend, that required him to more fully activate the non-judgmental model and apply it to the relevant situation. This caused him to blend with the model, taking its perspective as the truth. When that perspective was then propagated to the self-critical model, the easiest way for the mind to resolve the conflict was simply to alter the model producing the self-critical thoughts.

This seems like a straightforward instance of belief juxtaposition, and one where I ended up independently deriving something like UtEB's memory reconsolidation model: I too noted that the relevant belief structures need to be simultaneously activated in the right way to allow for the brain to revise one of them after noticing the contradiction. In general, UtEB's model of how things work rings true in my experience, making me inclined to believe that its description of how therapy works is correct, and that its model of how it is connected to neuroscience might also be.

UtEB and the subagent model

As many readers know, I have been writing a sequence of posts[16] on multi-agent models of mind. In Building up to an Internal

Family Systems model,[7] I suggested that the human mind might contain something like subagents which try to ensure that past catastrophes do not repeat. In *subagents*, coherence, and *akrasia in humans*,[17] I suggested that behaviors such as procrastination, indecision, and seemingly inconsistent behavior result from different subagents having disagreements over what to do.

As I already mentioned, my post on integrating disagreeing subagents[12] took the model in the direction of interpreting disagreeing subagents as conflicting beliefs or models within a person's brain. *Subagents, trauma and rationality*[18] further suggested that the appearance of drastically different personalities within a single person might result from unintegrated memory networks, which resist integration due to various traumatic experiences.

This post has discussed UtEB's model of conflicting emotional schemas in a way which further equates "subagents" with beliefs - in this case, the various schemas seem closely related to what e.g. Internal Family Systems calls "parts". In many situations, it is probably fair to say that this is what subagents are.

That said, I think that while this covers a very important subset of subagents, not everything which I have been referring to as a subagent falls straightforwardly under the belief-schema model. In *subagents* and *neural Turing machines*[19] as well as *Against "System 1" and "System 2"*,[20] I also covered subagents in a more general way, as also including e.g. the kinds of subsystems which carry out object recognition and are used to carry out tasks like arithmetic. This was also the lens through which I looked at subagents in my summary of Consciousness and the Brain.[13] Which kind of view is the most useful, depends on exactly what phenomenon we are trying to understand.

(1) *Bruce Ecker, Robin Ticic, Laurel Hulley, Robert A. Neimeyer. Unlocking the Emotional Brain: Eliminating Symptoms at Their Roots Using Memory Reconsolidation, 2012.*

(2) *LessWrong. How To Actually Change Your Mind.*

(3) *alkjash. Internal Double Crux, 2018.*

(4) *TurnTrout. Internalising Internal Double Crux, 2018.*

(5) *Jacob Falkovich. Internal Diet Crux, 2018.*

(6) *Duncan A Sabien. "Focusing" for skeptics, 2018.*

(7) *Kaj Sotala. Building up to an Internal Family Systems model, 2019.*

(8) *SaraHax. Actually updating, 2019.*

(9) *Eliezer Yudkowsky. Conservation of Expected Evidence, 2007.*

(10) *Ben Pace. A Sketch of Good Communication, 2018.*

(11) *Wikipedia. Overfitting.*

(12) *Kaj Sotala. Integrating disagreeing subagents, 2019.*

(13) *Kaj Sotala. Book Summary: Consciousness and the Brain, 2019.*

(14) *R. Lane, L. Ryan, L. Greenberg. Memory reconsolidation, emotional arousal, and the process of change in psychotherapy: New insights from brain science., 2015.*

(15) *Kaj Sotala. Tentatively considering emotional stories (IFS and "getting into Self"), 2018.*

(16) *Kaj Sotala. Multiagent Models of Mind, 2019.*

(17) *Kaj Sotala. Subagents, akrasia, and coherence in humans, 2019.*

(18) *Kaj Sotala. Subagents, trauma and rationality, 2019.*

(19) *Kaj Sotala. Subagents, neural Turing machines, thought selection, and blindspots, 2019.*

(20) *Kaj Sotala. System 2 as working-memory augmented System 1 reasoning, 2019.*

Raymond Arnold

Noticing Frame Differences

When disagreements persist despite lengthy good-faith communication, it may not just be about factual disagreements— it could be due to people operating in entirely different frames— different ways of seeing, thinking and/or communicating.

If you can't notice when this is happening, or you don't have the skills to navigate it, you may waste a lot of time.

Examples of Broad Frames

Gears-oriented Frames

Bob and Alice's conversation is about cause and effect. Neither of them are planning to take direct actions based on their conversation, they're each just interested in understanding a particular domain better.

Bob has a model of the domain that includes gears A, B, C and D. Alice has a model that includes gears C, D and F. They're able to exchange information, and their information is compatible, and they each end up with a shared model of how something works.

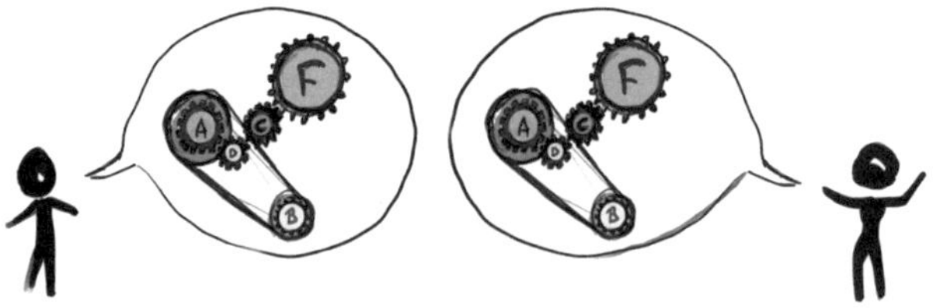

There are other ways this could have gone. Ben Pace covered some of them in a sketch of good communication[1]:

- Maybe they discover their models don't fit, and one of them is wrong
- Maybe combining their models results in a surprising, counter-intuitive outcome that takes them awhile to accept.
- Maybe they fail to integrate their models, because they were working at different levels of abstraction and didn't realize it. Sometimes they might fall into subtler traps.[2]

Maybe the thing Alice is calling "Gear C" is actually different from Bob's "Gear C". It turns out that they were using the same words to mean different things, and even though they'd both read blogposts warning them about that they didn't notice.

So Bob tries to slot Alice's gear F into his gear C and it doesn't fit. If he doesn't already have reason to trust Alice's epistemics, he may conclude Alice is crazy (instead of them referring to subtly different concepts).

This may cause confusion and distrust.

But, the point of this blogpost is that Alice and Bob have it easy.

They're actually trying to have the same conversation. They're both trying to exchange explicit models of cause-and-effect, and come away with a clearer understanding of the world through a reductionist lens.

There are many other frames for a conversation though.

Feelings-Oriented Frames

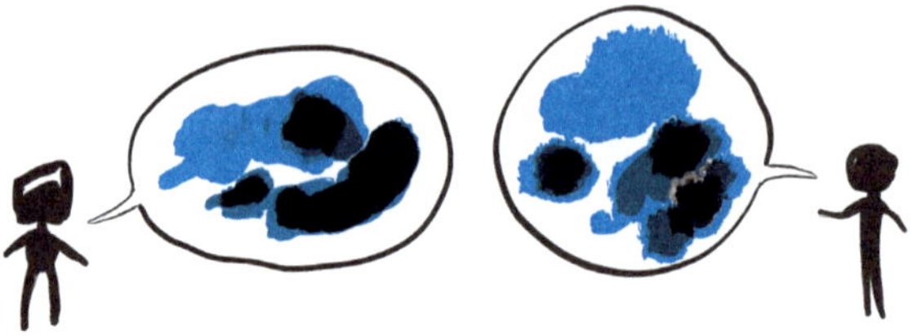

Clark and Dwight are exploring how they feel and relate to each other.

The focus of the conversation might be navigating their particular relationship, or helping Clark understand why he's been feeling frustrated lately.

When the Language of Feelings justifies itself to the Language of Gears, it might say things like: "Feelings are important information, even if it's fuzzy and hard to pin down or build explicit models out of. If you don't have a way to listen and make sense of that information, your model of the world is going to be impoverished. This involves sometimes looking at things through lenses other than what you can explicitly verbalize".

I think this is true, and important. The people who do their thinking through a gear-centric frame should be paying attention to feelings-centric frames for this reason. (And meanwhile, feelings themselves totally have gears that can be understood through a mechanistic framework).

But for many people that's not actually the point when looking through a feelings-centric frame. And not understanding this may lead to further disconnect if a Gearsy person and a Feelingsy person are trying to talk.

"Yeah feelings are information, but, also, like, man, you're a human being with all kinds of fascinating emotions that are an important part of who you are. This is super interesting! And there's a way of making sense of it that's necessarily experiential rather than about explicit, communicable knowledge".

Frames of Power and Negotiation

Dominance and Threat

Erica is Frank's boss. They're discussing whether the project Frank has been leading should continue, or whether it should stop and all the people on Frank's team should be reassigned.

Frank argues there's a bunch of reasons his project is important to the company (i.e. it provides financial value). He also argues that it's good for morale, and that cancelling the project would make his team feel alienated and disrespected.

Erica argues back that there are other projects that are more financially valuable, and that his team's feelings aren't important to the company.

It so happens that Frank had been up for a promotion soon, and that would put him (going forward) on more even footing with Erica, rather than her being his superior.

It's not (necessarily) about the facts, or feelings.

If Alice and Bob wandered by, they might notice Erica or Frank seeming to make somewhat basic reasoning mistakes about how much money the project would make or why it was valuable. Naively, Alice might point out that they seem to be engaging in motivated reasoning.

If Clark or Dwight wandered by, they might notice that Erica doesn't seem to really be engaging with Frank's worries about team morale. Naively, Clark might say something like "Hey, you don't seem to really be paying attention to what Frank's team is experiencing, and this is probably relevant to actually having the company be successful".

But the conversation is not about sharing models, and it's not about understanding feelings. It's not even necessarily about "what's best for the company".

Their conversation is a negotiation. For Erica and Frank, most of what's at stake are their own financial interests, and their social status within the company.

The discussion is a chess board. Financial models, worker morale, and explicit verbal arguments are more like game pieces than anything to be taken at face value.

This might be fully transparent to both Erica and Frank (such that neither even considers the other deceptive). Or, they might both earnestly believe what they're saying – but nonetheless, if you try to interpret the conversation as a practical decision about what's best for the company, you'll come away confused.

The Language of Trade

George and Hannah are negotiating a trade.

Like Erica and Frank, this is ultimately a conversation about what George and Hannah want.

A potential difference is that Erica and Frank might think of their situation as zero-sum, and therefore most of the resolution has more to do with figuring out "who would win in a political fight"?, and then having the counterfactual loser back down.

Whereas George/Hannah might be actively looking for positive sum trades, and in the event that they can't find one, they just go about their lives without getting in each other's way.

(Erica and Frank might also look for opportunities to trade, but doing so honestly might first require them to establish the degree to which their desires are mutually incompatible and who would win a dominance contest. Then, having established their respective positions, they might speak plainly about what they have to offer each other).

Noticing obvious frame differences

So the first skill here, is noticing when you're having wildly different expectations about what sort of conversation you're having.

If George is looking for a trade and Frank is looking for a fight, George might find himself suddenly bruised in ways he wasn't prepared for. And/or, Frank might have randomly destroyed resources when there'd been an opportunity for positive sum interaction.

Or: If Dwight says "I'm feeling so frustrated at work. My boss is constantly belittling me", and then Bob leaps in with an explanation of why his boss is doing that and maybe trying to fix it…

Well, this one is at least a stereotypical relationship failure mode you've probably heard of before (where Dwight might just want validation).

Untangling Emotions, Beliefs and Goals

A more interesting example of Gears-and-Feelings might be something like:

Alice and Dwight are talking about what career options Dwight should consider. (Dwight is currently an artist, not making much money, and has decided they want to try something else).

Alice says "Have you considered becoming a programmer? I hear they make a lot of money and you can get started with a 3 month bootcamp".

Dwight says "Gah, don't talk to me about programming".

It turns out that Dwight's dad always pushed him to learn programming, in a fairly authoritarian way. Now Dwight feels a bunch of ughiness around programming, with a mixture of "You're not the boss of me! I'mma be an artist instead"!

In this situation, perhaps the best option might be to say: "okay, seems like programming isn't a good fit for Dwight", and move on.

But it might also be that programming is actually a good option for Dwight to consider… it's just that the conversation can't proceed in the straightforward cost/benefit analysis frame that Alice was exploring.

Dwight making meaningful updates on whether programming is good for him depends on untangling his emotions, and/or exploring the relationship between his explicit models and his messier internals. It might require making peace with some longstanding issues with his father, or learning to detach them from the "should I be a programmer" question.

It might be that the most useful thing Alice can do is give him the space to work through that on his own.

If Dwight trusts Alice to shift into a feelings-oriented framework (or a framework that at least includes feeling), Alice might be able to directly help him with the process.

It may also be that this prerequisite trust doesn't exist, or that Dwight just doesn't want to have this conversation, in which case it's probably just best to move on to another topic.

Subtle differences between frames

This gets much more complicated when you observe that a) there's lots of slight variations on frames, and b) many people and conversations involve a mixture of frames.

It's not that hard to notice that one person is in a feelings-centric frame while another person is in a gears-centric frame. But things can actually get even more confusing if two people share a broad frame (and so think they should be speaking the same language), but actually they're communicating in two different subframes.

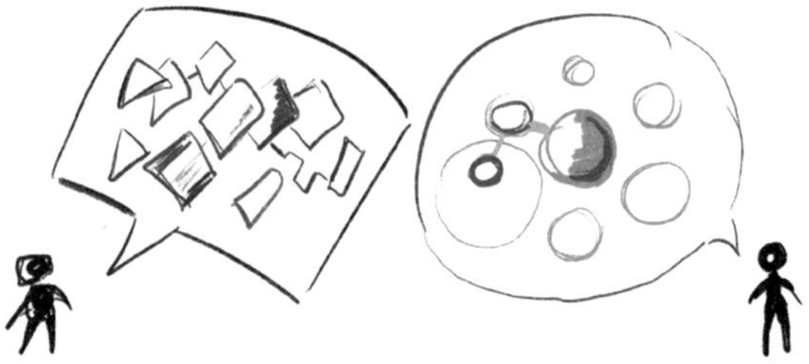

Example differences between gears-frames

Consider variations of Alice and Bob – both focused on causal models – who are coming from these different vantage points:

Goal-oriented vs Curiosity-driven conversation

Alice is trying to solve a specific problem (say, get a particular car engine fixed), and Bob thinks they're just, like, having a freewheeling conversation about car engines and how neat they are (and if their curiosity took them in a different direction they might shift the conversation towards something that had nothing to do with car engines).

Debate vs Doublecrux

Alice is trying to present arguments for her side, and expects Bob to refute those arguments or present different arguments. The burden of presenting a good case is on Bob.

Whereas Bob thinks they're trying to mutually converge on true beliefs (which might mean adopting totally new positions, and might involve each person focusing on how to change their own mind rather than their partner's).

Specific ontologies

If one person is, say, really into economics, then they might naturally frame everything in terms of transactions. Someone else might be really into programming and see everything as abstracted functions that call each other.

They might keep phrasing things in terms that fit their preferred ontology, and have a hard time parsing statements from a different ontology.

Example differences between feelings-frames

"Mutual Connection" vs "Turn Based Sharing"
Clark might be trying to share feelings for the sake of building connection (sharing back and forth, getting into a flow, getting resonance).

Whereas Dwight might think the point is more for each of them to fully share their own experience, while the other one listens and takes up as little space as possible.

"I Am My Feelings" vs "My Feelings are Objects"
Clark might highly self identify with his feelings (in a sort of Romantic framework). Dwight might care a lot about understanding his feelings but see them as temporary objects in his experience (sort of Buddhist).

Concrete example: The FOOM Debate

One of my original motivations for this post was the Yudkowsky/Hanson Foom Debate,[3] where much ink was spilled but AFAICT neither Yudkowsky nor Hanson changed their mind much.

I recently re-read through some portions of it. The debate seemed to feature several of the "differences within gears-orientation" listed above:

Specific ontologies: Hanson is steeped in economics and sees it as the obvious lens to look at AI, evolution and other major historical forces. Yudkowsky instead sees things through the lens of optimization, and how to develop a causal understanding of what recursive optimization means and where/whether we've seen it historically.

Goal vs Curiosity: I have an overall sense that Yudkowsky is more action oriented – he's specifically setting out to figure out the most important things to do to influence the far future. Whereas Hanson mostly seems to see his job as "be a professional economist, who looks at various situations through an economic lens, and see if that leads to interesting insights".

Discussion format: Throughout the discussion, Hanson and Yudkowsky are articulating their points using very different styles. On my recent read-through, I was impressed with the degree and manner to which they discussed this explicitly:

Eliezer notes:

> I think we ran into this same clash of styles last time (i.e., back at Oxford). I try to go through things systematically, locate any possible points of disagreement, resolve them, and continue. You seem to want to jump directly to the disagreement and then work backward to find the differing premises. I worry that this puts things in a more disagreeable state of mind, as it were—conducive to feed-backward reasoning (rationalization) instead of feed-forward reasoning.
>
> It's probably also worth bearing in mind that these kinds of metadiscussions are important, since this is something of a trailblazing case here. And that if we really want to set up conditions where we can't agree to disagree, that might imply setting up things in a different fashion than the usual Internet debates.

Hanson responds:

> When I attend a talk, I don't immediately jump on anything a speaker says that sounds questionable. I wait until they actually make a main point of their talk, and then I only jump on points that seem to matter for that main point. Since most things people say actually don't matter for their main point, I find this to be a very useful strategy. I will be very surprised indeed if everything you've said mattered regarding our main point of disagreement.

I found it interesting that I find both these points quite important – I've run into each failure mode before. I'm unsure how to navigate between this rock and hard place.[4]

• • •

My main goal with this essay was to establish frame-differences as an important thing to look out for, and to describe the concept from enough different angles to (hopefully) give you a general sense of what to look for, rather than a single failure mode.

What to do once you notice a frame-difference depends a lot on context, and unfortunately I'm often unsure what the best approach is. The next few posts will approach "what has sometimes worked for me", and (perhaps more sadly) "what hasn't".

(1) *Ben Pace. A Sketch of Good Communication. 31st March 2019.*

(2) *Eliezer Yudkowsky. Making Beliefs Pay Rent (in Anticipated Experiences). 28th July 2007.*

(3) *Robin Hanson & Eliezer Yudkowsky. The Hanson-Yudkowsky AI-FOOM Debate. 2008.*

(4) *Ruby Bloom. Do you fear the rock or the hard place? 20th July 2019.*

Raymond Arnold

Propagating Facts into Aesthetics

Epistemic status: Tentative. I've been practicing this on-and-off for a year and it's seemed valuable, but it's the sort of thing I might look back on and say "hmm, that wasn't really the right frame to approach it from".

In doublecrux, the focus is on "what observations would change my mind"?

In some cases this is (relatively) straightforward. If you believe minimum wage helps workers, or harms them, there are some fairly obvious experiments you might run. "Which places have instituted minimum wage laws? What happened to wages? What happened to unemployment? What happened to worker migration"?

The details will matter a lot. The results of the experiment might be weird and confusing.[1] If I ran the experiment myself I'd probably get a lot of things wrong, misuse statistics and forget to account for some confounding factors. But I don't feel confused about how to learn better statistics, account for more confounders, etc.

But there's a problem that seems harder to me, which is *how to change my mind about aesthetics*. Sarah Constantin first brought this up in Naming the Nameless,[2] and I've been thinking about it ever since.

I think a lot of deep disagreements have to do with "what is beautiful, and what is ugly"?, and inability to directly address this is part of what prevents those disagreements from resolving.

In the case of the minimum wage example, you might run an experiment, and find overwhelming evidence that minimum wage helps or hurts workers. But because there's lots of confounders, the evidence might be mixed and confusing. How you interpret it will depend on how it fits into your existing worldview.

Part of this has to do with your ontological frame. But I think a lot has to do with aesthetics judgments, such as:

- *Is capitalism ugly and/or distasteful?* You might have very salient examples of how capitalism can result in exploitation, pollution, or people becoming trapped in unhealthy power structures.
- *Is capitalism beautiful?* Alternately, it might be salient that capitalism creates supermarkets, gains from trade, and vast surplus. Economic efficiency isn't just pretty numbers on a graph, it's real value being created.

These don't directly bear on the minimum wage question, but might make it harder to resolve.

In some cases, your aesthetic taste might make it harder to update on new information properly. In other cases, your aesthetic taste might help you to notice important patterns more readily. I'm using the word aesthetic in a nonstandard way. When people do that, I think it's important to be clear and about what they're doing and why.

Why 'aesthetics'?

There's a few different words I might have used here, including "feelings", "ontologies", "frameworks", and "values".

Most obviously, I could have asked 'is capitalism good/bad'? instead of 'is capitalism beautiful or ugly'?

I'm making a fairly strong claim (weakly held) that "is it beautiful or ugly"? is at least one of the important questions to be asking, in addition to "is capitalism *good/bad*" and "does raising minimum wage help or harm workers"? Not because it's how a flawless AI would think about it, but because it's how humans seem to often think about it.

What is an aesthetic?

An aesthetic is a mishmash of values, strategies, and ontologies that reinforce each other.

The values reinforce "you want to use strategies that achieve these values".

The act of using a particular strategy shapes the ontology that you see the world through.[3]

The ontology reinforces what values seem important to you.

Together, this all creates a feedback loop between your metagoals and subgoals, where the process of using this cluster of value/strategy/ontology makes each link in the chain stronger.

In humans (who have messy, entangled brains), this cashes out into feelings, felt senses. The original goal and the metagoals blur together. I think "this helps me achieve my [generic] goals" might reinforce "these *particular* subgoals I have are good goals to help with my overall flourishing".

This might be implemented via evolution over millions of years, or via human brains over decades. A Just So Story I'm not sure I endorse but hopefully gets the point across[4]:

> "A flower is beautiful, you say. Do you think there is no story behind that beauty, or that science does not know the story? Flower pollen is transmitted by bees, so by sexual selection, flowers evolved to attract bees—by imitating certain mating signs of bees, as it happened; the flowers' patterns would look more intricate, if you could see in the ultraviolet. Now healthy flowers are a sign of fertile land, likely to bear fruits and other treasures, and probably prey animals as well; so is it any wonder that humans evolved to be attracted to flowers"?

Here are some things that you might find beautiful, or distasteful:

- Mozart
- Punk Rock
- Readable, well-written code
- Clever hacks that got the job done quickly
- Cities built on rectangular grids[5]
- Winding alleyways in villages where nobody has consistent names[6]
- People being physically affectionate in public
- A harsh, barren desert
- A lush valley with a river
- Swamps/ wetlands
- Nature in general
- Manicured gardens

- Books
- Throwing away books
- People speaking in languages different from yours
- Dense spreadsheets laden with accurate data
- Minimalism
- Frugalism
- Patriotism
- People going out of their way to be kind to their neighbors
- People going out of their way to solve small-but-common problems using math

(If you're like me, you might find it distasteful when people make moral arguments that seem rooted in distaste... and then feel kinda self conscious about the contradiction).

Sometimes you're doublecruxing with someone, and they've explained their model. And their model… makes sense. But the conclusion just seems *so damn ugly*. You want to take 5x the time to write beautiful code, and they just want you to get the job done and ship it.

One thing you can do is push aside your aesthetic judgment, shut up and multiply. This may be useful for expediency.

But sometimes, I think the correct thing is for one or both people to backpropagate facts through their aesthetics.

I do *not* think you should rush or "force" this. Your sense of beauty is there for a reason. But I have a sense that figuring out how to do this well is a key open problem in applied rationality.

Examples

Are Swamps Beautiful?

Compare the swamp with a verdant forest.

If you're like me, swamps seem ugly. Forests seem pretty.

My associations with swamps come largely from stories (and perhaps most concretely, from the game "Magic the Gathering"), where they're often presented as places of disease, murky horrors and corrupt magic. In person, swamps are physically hard to walk in (sometimes solid ground turns out to be algae), and full of mosquitoes that bite me.

Are these associations accurate?

Well, the solid ground and mosquito issues are definitely real.

Swamp Thing[7] is not real, life-stealing magic[8] is not real.

Are there additional facts I can learn? My sister evaluates land for construction projects. She says that swamps often serve important roles as a natural way to filter water, and when you naively drain swamps, water quality in an area gets worse. James Scott in Against the Grain[9] claims that early Sumerian civilization developed in swamps, where food and resources were plentiful and life was fairly leisureful – until empires arose and subjected people and forced them to switch to easily-taxable crops instead of the ones that grew naturally. (Speaking of which: is civilization beautiful or ugly)?

I probably find forests beautiful, in part, because they represent a lot of resources that I understand how to make use of. If swamps also supply those resources, maybe I should respect them more?

I also find forests beautiful because my experience stems from a) enchanted forests in fairytales, and b) relatively manicured national parks. If I remind myself that the last time I walked through an *untamed* forest, it was dense with brambles that cut me 'til I bled. It wasn't actually a much nicer experience than the last time I explored a swamp. (It also had non-trivial numbers of mosquitoes).

In this example, simply mulling over the facts naturally re-organizes my feelings about them. I still find swamps ugly, but less ugly than before. I expect that, if I reflected on this periodically, over time, it would shift a bit more.

Are Harsh Deserts Beautiful?

I am in fact confused by this. My answer is "yes", and I don't know why. Deserts don't have much in the way of resources. Their stark beauty is more like the way a statue is beautiful than the way a forest is beautiful.

I mulled this one over for a while, am still confused and I note it here because "noticing the limits of a model" seems important.

[Edit: this was discussed more in the comments].[10]

Is Helping Nearby Other People Beautiful?

The first experience I got with aesthetic doublecrux was debating "Hufflepuff virtue" with Oliver Habryka.

I had a strong sense that "people helping each other out" was good and right and virtuous. There was a beauty to the sort of community where everyone notices when someone is hurting (and reaches out to help), or when a space is messy (and cleans it up).

There was a cluster of attributes that seemed to fit together in a way that was stronger than the sum of its parts.

And this was visibly lacking in the Berkeley community, and it was resulting in people feeling alienated and distrustful of each other, and many spaces being either messy, or burdening a single person with cleaning up everyone else's mess.

This seemed concretely harmful. But it also just seemed...
ugly and bad.

Oliver had a different view, which I summarize as the "systemization and specialization" approach.[11]

If everyone has to pay attention to their environment and notice things that need doing, this is a *lot* of cognitive overhead. If people only have seven working memory slots but they're spending one of them on tracking the environment, that's a dramatic cost on their ability to think. For a community that specializes in thinking, this could be quite bad.

Moreover, "everyone pitch in" is just a really inefficient way of getting things done. A better solution is to streamline and automate as much of the work as possible, hire cleaning services, and whatever remaining work needs doing, simply pay one one of the people something commensurate for their time and effort. Specialization is how things get done when you're doing them seriously.

We argued about this over the course of three days.

I still think there are some things Habryka was missing here. But eventually my worldview shifted in some significant ways:

- I updated that the "everyone pitch in" way of keeping spaces clean doesn't make sense for longterm organizations with serious funding. Specialization is real, cognitive bandwidth is precious, and it's generally better to just hire a cleaning service if you can afford one.
- I updated a bit (talking with Satvik) that my model that "helping each other out in low-key ways builds trust which later enables more extensive projects" wasn't as strong as I thought. Satvik asked something like "do you think startup cofounders tend to team up because they've helped each other take out the trash? I feel like it's more about sharing a clear vision and principles or something". And I thought back to some experiences and... yeah that seemed maybe more accurate.
- I gained a better understanding of where and why the "everyone pitch in" approach is useful.
    - Cleaning services are expensive, and if you're a fledgling organization or a typical household, it's probably not worth hiring a cleaner more than once a week or so. Meanwhile, people make messes much more frequently than once a week. If you want your space nice, you have to clean it yourself.
    - There's a value that comes from having community spaces use the "everyone pitch in" method, in that it creates a stronger sense of ownership and buy-in for the space. It also is a mechanism by which people can relate to each other more easily. While this might not be that important for a company, it seems important for a community that's aiming to meet community-shaped-needs.

But this all left me with a nagging, frustrated sense that something important and beautiful was being lost. I want to live in a world where people help each other out in small ways. It's the particular kind of beauty that a small town in a Miyazaki movie embodies. It feels important to me.

Under what circumstances should I change how I feel about that?

There's a sense in which aesthetics can't be proven wrong, or at least "trying to prove it wrong" isn't really the right frame of mind.

But… I have an aesthetic preference for consistency, and for believing true things (whether this is good is another question, but I'm taking it at face value for now), which informs my other aesthetics. Aesthetics can turn out to be built out of contradictory pieces, and they can turn out to hinge on false beliefs.

"Trying on" another aesthetic

While talking to Habryka, I tried to get a sense of what it's like to live in the world where systemization and specialization are obviously good and right. What was it like to be Habryka? How did this fit together with his other beliefs and values?

Then, once I had a good handle on that, I tried to inhabit "what would it be like to be a Raemon who found systemization and specialization good and right?". Without actually adopting the aesthetic, I tried fitting it into my existing model. This was a bit of an aesthetic process of its own – like trying on a new outfit and seeing how I reacted in the mirror.

I'm not sure if Habryka endorses considering those as an 'aesthetic', per se. But I found this process valuable.

I gained some ability to see systemization as beautiful. My sense of Hufflepuff beauty became more nuanced and caveated.

Clean Code vs Quick Hacks

Humans have (an instinctive? Learned? I'm not sure) sense that when you smell fecal matter or rotting flesh, there is probably disease nearby. It's digusting.

Dogs... well, I'm not 100% sure what's going on with dogs but I think it's something like "strong odors that mask my scent are more useful than disease is bad", and for some reason fecal matter is joyful to play around in.

Programmers often learn that spaghetti code is evidence of bugs, even if they don't know exactly what the bug is yet. It acquires a *bad code smell.*

Young programmers often do not have this sense of distaste, and it is important for them to acquire it.

On the flipside: there is also a thing where, well, sometimes you're rushing to ship a Minimum Viable Product and you don't have time to do everything right. It can be legitimately hard to figure out how much effort to put into "doing things right". But it seems at least sometimes, experienced coders either need to learn to "hold their nose" and do the quick fix, or to develop alternate aesthetics that they can shift between depending on circumstances.

Knobs to Turn

There are a few different directions this kind of process might go:

- You could shift to find something more beautiful than you did before
- You could shift to find something less beautiful than you did before
- You could shift to find something more distasteful than you did before
- You could shift to find something less distasteful than you did before

I have some sense that these are subtly different processes, although not much evidence to back that up. I also feel like in each case, going from Zero to N, or N to Zero, is different than dialing an existing aesthetic response up or down.

Gaining a new appreciation for why something is beautiful feels different than gaining a categorically new form of disgust. In particular, gaining a new form of beauty mostly makes my life feel nicer, whereas gaining a new form of disgust increases the unpleasantness.

Why Does this Matter?

In Naming the Nameless,[12] Sarah Constantin references this comment by Scott Alexander[13]:

> Sometimes I can almost feel this happening. First I believe something is true, and say so. Then I realize it's considered low-status and cringeworthy. Then I make a principled decision to avoid saying it – or say it only in a very careful way – in order to protect my reputation and ability to participate in society. Then when other people say it, I start looking down on them for being bad at public relations. Then I start looking down on them just for being low-status or cringeworthy.

> Finally the idea of "low-status" and "bad and wrong" have merged so fully in my mind that the idea seems terrible and ridiculous to me, and I only remember it's true if I force myself to explicitly consider the question. And even then, it's in a condescending way, where I feel like the people who say it's true deserve low status for not being smart enough to remember not to say it. This is endemic, and I try to quash it when I notice it, but I don't know how many times it's slipped my notice all the way to the point where I can no longer remember the truth of the original statement".

Sarah notes:

> Now, I could say "just don't do that, then" – but Scott of 2009 would have also said he believed in being independent and rational and not succumbing to social pressure. Good intentions aren't enough. [...]
>
> I think it's much better to try to make the implicit explicit, to bring cultural dynamics into the light and understand how they work, rather than to hide from them.

Scott's comment gets at what I mean by "An aesthetic is a mishmash of values, strategies, beliefs, and ontologies that reinforce each other". He starts with a belief, then adopts a strategy for how he relates his communication to that belief, and then ends up with a vague sense that the belief is "cringey", and later collapsing it to "cringey and wrong".

This is quite worrying epistemic horror.

I think most of what needed saying, Sarah already said, but it's worth concluding with here:

> If you take something about yourself that's "cringeworthy" and, instead of cringing yourself, try to look at why it's cringeworthy, what that's made of, and dialogue honestly with the perspective that disagrees with you – then there is, in a sense, nothing to fear.
>
> There's an "elucidating" move that I'm trying to point out here, where instead of defending against an allegation, you say "let's back up a second" and bring the entire situation into view. It's what double crux[14] is about – "hey, let's find out what even is the disagreement between us." Double crux is hard enough with arguments, and here I'm trying to advocate something like double-cruxing aesthetic preferences, which sounds absurdly ambitious. But: imagine if we could talk about why things seem beautiful and appealing, or ugly and unappealing. Where do these preferences come from, in a causal sense? Do we still endorse them when we know their origins? What happens when we *bring tacit things into consciousness,* when we talk carefully about what aesthetics evoke in us, and how that might be the same or different from person to person?
>
> Unless you can think about how cultural messaging works, you're going to be a *mere* consumer of culture, drifting in whatever direction the current takes you.

I'm hoping this post gives some nuts and bolts on how to actually make progress on that goal.

Again, I don't know that the specific techniques I list in this post are the best ones, or how often exactly aesthetic concerns are most relevant. I think it's usually good form to start with an attempt to take arguments at face value, and debate about concrete beliefs.

But, if that isn't working, I think digging into aesthetics is one of the tools that's important to have in your toolkit.

 Originally published 18th December 2019

(1) *Zvi Mowshowitz. On the Seattle Minimum Wage Study (part 1), 2017.*

(2) *Sarah Constantin. Naming the Nameless, 2018.*

(3) *Raymond Arnold. Picture Frames, Window Frames and Frameworks, 2019.*

(4) *Eliezer Yudkowsky. The Gift We Give To Tomorrow. 2008.*

(5) *Scott Alexander. Book Review: Seeing Like a State. 2017.*

(6) *Scott Alexander. Book Review: Seeing Like a State. 2017.*

(7) *Swamp Thing, DC Comics Universe TV Series. 2019.*

(8) *Magic the Gathering, Corrupt Card.*

(9) *James C. Scott. Against the Grain: A Deep History of the Earliest States. 2017.*

(10) *Comments on Propagating Facts into Aesthetics, Lesswrong.com. 2019.*

(11) *Raymond Arnold. Hufflepuff Leadership and Fighting Entropy. 2018.*

(12) *Sarah Constantin. Naming the Nameless, 2018.*

(13) *Scott Alexander. Does Age Bring Wisdom? 2017.*

(14) *Duncan Sabien. Double Crux—A Strategy for Resolving Disagreement. 2017.*

Ruben Bloom

Do you fear the rock or the hard place?

Epistemic status: fairly confident based on my accumulated experience of debates and disagreements. I wrote this for myself as much as others.

There is a conversational dynamic which I think is extremely common, a failure mode which is all too easy to fall into. Alice and Bob are debating some course of action, e.g. should they do X or Y? Alice thinks that X is very likely to result in terrible consequence R, so they should definitely opt for Y. Bob thinks that Y most definitely will cause horrific result H, so they should definitely do X.

The distilled conversation goes a bit like this:

> Alice: "We can't do X! That would lead to R, which is unacceptable".

Bob: "I don't think you get it, Y results in H. You can't think that we could allow H, do you"?

Alice: "I feel like you're not listening, we need to account for R"!

Bob: "H is definitely a much worse and more real danger than R"...

Alice is afraid of **the rock** (R) and Bob is afraid of **the hard place** (H).

Possible values of X, Y, R, and H:

X = more gun control; Y = less gun control; R = people unable to defend themselves and having their rights taken away; H = increased risk of mass shootings, suicides, and children shooting themselves or others.

X = raising minimum wage; Y = maintaining minimum wage; R = reduction in number of jobs causing people to be fired; H = people not earning enough from their jobs to subsist.

X = increase immigration; Y = reduce immigration; R = loss of jobs from local community, erosion of national culture and values, crime committed by migrants; H = humanitarian impact, loss of potential growth of the national economy.

The above exchange is actually relatively good. Alice and Bob each know what they're afraid of and have expressed that clearly. Bob even acknowledges Alice's concern about R, but states that he thinks it's the lesser danger. They're at a point where they might be able to usefully double crux.[1]

What goes wrong?

Failure to identify and articulate the fears

If Carol has held the position that X is really bad for a long time, or if her position stems from deep System 1 models and frames, then she might struggle to articulate clearly what specifically she's afraid that X will cause. She might find any attempts by others to clarify to be unsatisfying, and possibly threatening because any incorrect articulation of your fear is often worse than none at all. Dylan might come along and say "you don't like X because you're afraid of P, but P won't happen, so you should be okay with X". This could be scary to Carol who feels her fears are just being played down so they can be dismissed.

If both Carol and Dylan are unable to voice what they're afraid of, the resulting conversation can be Carol and Dylan simply shouting each other about how terrible and evil they think the other's position is. It becomes one person's cached store of fear and horror pitted against another's.

Failure to acknowledge the other person's fear while being desperate for yours to be acknowledged

Caught up in her dread of R, Alice can become insistent that Bob acknowledges the extreme danger she sees. Bob failing to do so is scary - perhaps he will advocate for X not realizing the tremendous harm he will cause.

Alice's fear of Bob's position can be overriding. It's easy for her to feel the conversation can't proceed until Bob can be made to realize what his position will result in. Indeed, if Bob can't see the extreme danger of X leading to R, then possibly he can't be reasoned with at all. Alice will focus all her attention, energy, and emotion on trying to make Bob see reason here.

This is not conducive to Alice listening to Bob. If Bob isn't acknowledging R, then it's easy to see all his words as a perverse and willful refusal to acknowledge R.

But it's worse! Bob is in exactly the same state as Alice. He is dreadfully afraid that Alice isn't worried about H. That she'd gladly and foolishly let H happen to avoid R. Does she just not care about H happening? Will she sacrifice it all so readily? How can he debate with someone with such distorted values? Someone who keeps ignoring his outright statements that Y leads to H!

You easily get two people yelling their fears at each other, unwilling to listen until the other person acknowledges the badness they have been advocating for. The conversation goes nowhere. Tones start out calm and civil, but rapidly become outraged at the wanton obtuseness of their infuriating interlocutor.

Refusal to acknowledge the other person's fear because Arguments Are Soldiers

> "Politics is the mind-killer. Arguments are soldiers. Once you know which side you're on, you must support all arguments of that side, and attack all arguments that appear to favor the enemy side; otherwise it's like stabbing your soldiers in the back. If you abide within that pattern, policy debates will also appear one-sided to you—the costs and drawbacks of your favored policy are enemy soldiers, to be attacked by any means necessary". - Policy Debates Should Not Appear One-Sided[2]

Even if you understand what your interlocutor is afraid of and you think there's something to it, it can be tempting to not acknowledge this. Acknowledging their fear can feel like putting points on the board for them and their position. So you deny them this, not willing to cede any ground.

This is bad, don't do it kids. It's possible that your concern is by far the greater one, but that doesn't mean their worries aren't legitimate too. (Or maybe they're not legitimate, but you can at least show you understand what they are feeling). The discussion can proceed if you at least acknowledge the fact they have their own fear too.

What to do?

If you think you might be headed for a Rock vs Hard Place dynamic, I'd suggest trying the following:

1. Understand what *you* are afraid of. What is the outcome you think is at risk of happening?

2. Understand what *they* are afraid of. Know what their fear is.

3. Acknowledge their fear. Acknowledge that something they think is very important is at stake in your discussion.
[See this important comment by Kaj on what this requires][3]
- Step 1 is simply to acknowledge the fear. The following steps depend on your own beliefs.
- You might say "I too am afraid of R", but:
    - I have thought about this and believe that X won't cause are R / can be made to avoid R.
    - I think that the danger of R is outweighed by the worse danger of H.

    - I think that there are ways to minimize the risk or damage that are good enough.
- You might say "I don't share your fear of R", because
    - I think you are mistaken that R is in fact bad.

4. Don't fall into Arguments Are Soldiers/Policy Debates are One-Sided traps.

5. Realize that the fear most salient to you will derive from your experience, background, situation, and models. Someone might reasonably be afraid in the other direction, and their situation has made that much clearer to them. There's a chance you can learn from them.

These steps can be taken somewhat unilaterally. You can acknowledge someone else's fear without them acknowledging yours. Possibly after you've done so, they can relax a little and will be open to acknowledging that you have your own fear.

If someone is completely unwilling to consider that their fear is misplaced, mistaken, or outweighed, then the conversation may struggle to go anywhere. Best to realize this quickly and find an alternative path forward than end up in circular conversations that only generate fear and hostility.

That said, I'm hopeful that once people are able to acknowledge their own and others fears, productive double cruxing can happen about where the balance lies between the rock and the hard place.

Bah!

Appendix: On a more positive note

While I've framed this post in terms of fears and potentially negative outcomes, it applies also in cases where people are disagreeing about different policies that would result in different benefits. Just as gradient ascent and gradient descent are effectively the same things, sometimes people are fighting over their fears of not netting some positive benefit.

Double-Crux Resources:

- Double Crux — A Strategy for Resolving Disagreement[1]
- Musings on Double Crux (and "Productive Disagreement")[4]
- A Concrete Multi-Step Variant of Double Crux I Have Used Successfully[5]
- *Another* Double Crux Framework[6]

(1) *Duncan Sabien. Double Crux — A Strategy for Resolving Disagreement. 1st Jan 2017.*

(2) *Eliezer Yudkowsky. Policy Debates Should Not Appear One-Sided. 3rd Mar 2007.*

(3) *Kaj Sotala (comment). Do you fear the rock or the hard place?. 21st Jul 2019.*

(4) *Raymond Arnold. Musings on Double Crux (and "Productive Disagreement"). 28th Sep 2017.*

(5) *deluks917. A Concrete Multi-Step Variant of Double Crux I Have Used Successfully. 14th Mar 2018.*

(6) *Raymond Arnold. *Another* Double Crux Framework. 28th May 2018.*

Scott Alexander

Mental Mountains

I.

Kaj Sotala has an outstanding review[1] of *Unlocking The Emotional Brain*;[2] I read the book, and Kaj's review is better.

He begins:

> UtEB's premise is that much if not most of our behavior is driven by emotional learning. Intense emotions generate unconscious predictive models of how the world functions and what caused those emotions to occur. The brain then uses those models to guide our future behavior. Emotional issues and seemingly irrational behaviors are generated from implicit world-models (schemas) which have been formed in response

to various external challenges. Each schema contains memories relating to times when the challenge has been encountered and mental structures describing both the problem and a solution to it.

So in one of the book's example cases, a man named Richard sought help for trouble speaking up at work. He would have good ideas during meetings, but felt inexplicably afraid to voice them. During therapy, he described his narcissistic father, who was always mouthing off about everything. Everyone hated his father for being a fool who wouldn't shut up. The therapist conjectured that young Richard observed this and formed a predictive model, something like "talking makes people hate you". This was overly general: talking only makes people hate you if you talk incessantly about really stupid things. But when you're a kid you don't have much data, so you end up generalizing a lot from the few examples you have.

When Richard started therapy, he didn't consciously understand any of this. He just felt emotions (anxiety) at the thought of voicing his opinion. The predictive model output the anxiety, using reasoning like "if you talk, people will hate you, and the prospect of being hated should make you anxious – therefore, anxiety", but not any of the intermediate steps. The therapist helped Richard tease out the underlying model, and at the end of the session Richard agreed that his symptoms were related to his experience of his father. But knowing this changed nothing; Richard felt as anxious as ever.

Predictions like "speaking up leads to being hated" are special kinds of emotional memory. You can rationally understand that the prediction is no longer useful, but that doesn't really help; the emotional memory is still there, guiding your unconscious predictions. What should the therapist do?

Here *UtEB* dives into the science on memory reconsolidation.

Scientists have known for a while that giving rats the protein synthesis inhibitor anisomycin prevents them from forming emotional memories. You can usually give a rat noise-phobia by pairing a certain noise with electric shocks, but this doesn't work if the rats are on anisomycin first. Probably this means that some kind of protein synthesis is involved in memory. So far, so plausible.

A 2000 study[3] found that anisomycin could also erase existing phobias in a very specific situation. You had to "activate" the phobia – get the rats thinking about it really hard, maybe by playing the scary noise all the time – and then give them the anisomycin. This suggested that when the memory got activated, it somehow "came loose", and the brain needed to do some protein synthesis to put it back together again.

Thus the idea of memory reconsolidation: you form a consolidated memory, but every time you activate it, you need to reconsolidate it. If the reconsolidation fails, you lose the memory, or you get a slightly different memory, or something like that. If you could disrupt emotional memories like "speaking out makes you hated" while they're still reconsolidating, maybe you could do something about this.

Anisomycin is pretty toxic, so that's out. Other protein synthesis inhibitors are also toxic – it turns out proteins are kind of important for life – so they're out too. Electroconvulsive therapy actually seems to work pretty well for this[4] – the shock disrupts protein formation very effectively (and the more I think about this, the more implications it seems to have). But we can't do ECT on everybody who wants to be able to speak up at work more, so that's also out. And the simplest solution –

activating a memory and then reminding the patient that they don't rationally believe it's true – doesn't seem to help; the emotional brain doesn't speak Rationalese.

The authors of *UtEB* claim to have found a therapy-based method that works, which goes like this:

First, they tease out the exact predictive model and emotional memory behind the symptom (in Richard's case, the narrative where his father talked too much and ended up universally hated, and so if Richard talks at all, he too will be universally hated). Then they try to get this as far into conscious awareness as possible (or, if you prefer, have consciousness dig as deep into the emotional schema as possible). They call this "the pro-symptom position" – giving the symptom as much room as possible to state its case without rejecting it. So for example, Richard's therapist tried to get Richard to explain his unconscious pro-symptom reasoning as convincingly as possible: "My father was really into talking, and everybody hated him. This proves that if I speak up at work, people will hate me too". She even asked Richard to put this statement on an index card, review it every day, and bask in its compellingness. She asked Richard to imagine getting up to speak, and feeling exactly how anxious it made him, while reviewing to himself that the anxiety felt justified given what happened with his father. The goal was to establish a wide, well-trod road from consciousness to the emotional memory.

Next, they try to find *a lived and felt experience* that contradicts the model. Again, Rationalese doesn't work; the emotional brain will just ignore it. But it will listen to experiences. For Richard, this was a time when he was at a meeting, had a great idea, but didn't speak up. A coworker had the same idea, mentioned it, and everyone agreed it was great, and congratulated the other person for having such an amazing idea that would transform their business. Again, there's this same process of trying to get as much in that moment as possible, bring the relevant feelings back again and again, create as wide and smooth a road from consciousness to the experience as possible.

Finally, the therapist activates the disruptive emotional schema, and before it can reconsolidate, smashes it into the new experience. So Richard's therapist makes use of the big wide road Richard built that let him fully experience his fear of speaking up, and asks Richard to get into that frame of mind (activate the fear-of-speaking schema). Then she asks him, *while keeping the fear-of-speaking schema in mind*, to remember the contradictory experience (coworker speaks up and is praised). Then the therapist vividly describes the juxtaposition while Richard tries to hold both in his mind at once.

And then Richard was instantly cured, and never had any problems speaking up at work again. His coworkers all applauded, and became psychotherapists that very day. An eagle named "Psychodynamic Approach" flew into the clinic and perched atop the APA logo and shed a single tear. *Coherence Therapy: Practice Manual And Training Guide* was read several times, and God Himself showed up and enacted PsyD prescribing across the country. All the cognitive-behavioralists died of schizophrenia and were thrown in the lake of fire for all eternity.

This is, after all, a therapy book.[5]

II.

I like *UtEB* because it reframes historical/purposeful[6] accounts of symptoms as aspects of a predictive model. We already know the brain has an unconscious predictive model that it uses to figure out how to respond to various situations and which actions have which consequences. In retrospect, this framing perfectly fits the idea of traumatic experiences having outsized effects. Tack on a bit about how the model is more easily updated in childhood (because you've seen fewer other things, so your priors are weaker), and you've gone a lot of the way to traditional models of therapy.

But I also like it because it helps me think about the idea of separation/noncoherence in the brain. Richard had his schema about how speaking up makes people hate you. He also had lots of evidence that this wasn't true, both rationally (his understanding that his symptoms were counterproductive) and experientially (his story about a coworker proposing an idea and being accepted). But the evidence failed to naturally propagate; it didn't connect to the schema that it should have updated. Only after the therapist forced the connection did the information go through. Again, all of this should have been obvious – of course evidence doesn't propagate through the brain, I was writing posts ten years ago about how even a person who knows ghosts don't exist will be afraid to stay in an old supposedly-haunted mansion at night with the lights off. But UtEB's framework helps snap some of this into place.

UtEB's brain is a mountainous landscape, with fertile valleys separated by towering peaks. Some memories (or pieces of your predictive model, or whatever) live in each valley. But they can't talk to each other. The passes are narrow and treacherous. They go on believing their own thing, unconstrained by conclusions reached elsewhere.

Consciousness is a capital city on a wide plain. When it needs the information stored in a particular valley, it sends messengers over the passes. These messengers are good enough, but they carry letters, not weighty tomes. Their bandwidth is atrocious; often they can only convey what the valley-dwellers think, and not why. And if a valley gets something wrong, lapses into heresy, as often as not the messengers can't bring the kind of information that might change their mind.

Links between the capital and the valleys may be tenuous, but valley-to-valley trade is almost non-existent. You can have two valleys full of people working on the same problem, for years, and they will basically never talk.

Sometimes, when it's very important, the king can order a road built. The passes get cleared out, high-bandwidth communication to a particular valley becomes possible. If he does this to two valleys at once, then they may even be able to share notes directly, each passing through the capital to get to each other. But it isn't the norm. You have to really be trying.

This ended out a little more flowery than I expected, but I didn't start thinking this way because it was poetic. I started thinking this way because of this:

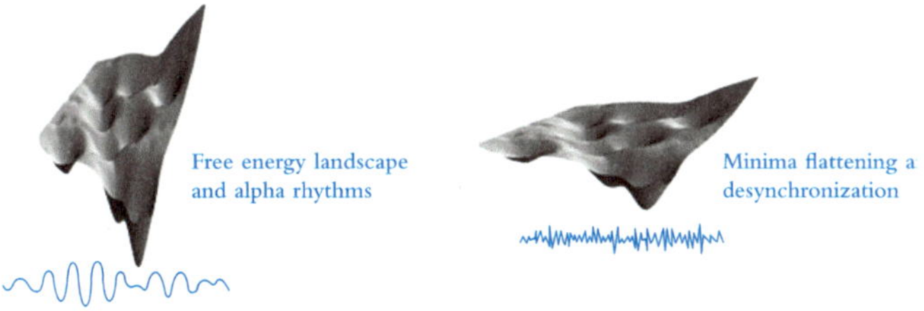

Frequent SSC readers will recognize this as from Figure 1 of Friston and Carhart-Harris' REBUS And The Anarchic Brain: Toward A Unified Model Of The Brain Action Of Psychedelics,[7] which I review here.[8] The paper describes it as "the curvature of the free-energy landscape that contains neuronal dynamics. Effectively, this can be thought of as a flattening of local minima, enabling neuronal dynamics to escape their basins of attraction and—when in flat minima—express long-range correlations and desynchronized activity".

Moving back a step: the paper is trying to explain what psychedelics do to the brain. It theorizes that they weaken high-level priors (in this case, you can think of these as the tendency to fit everything to an existing narrative), allowing things to be seen more as they are:

> A corollary of relaxing high-level priors or beliefs under psychedelics is that ascending prediction errors from lower levels of the system (that are ordinarily unable to update beliefs due to the top-down suppressive influence of heavily-weighted priors) can find freer register in conscious experience, by reaching and impressing on higher levels of the hierarchy. In this work, we propose that this straightforward model can account for the full breadth of subjective phenomena associated with the psychedelic experience.

These ascending prediction errors (ie noticing that you're wrong about something) can then correct the high-level priors (ie change the narratives you tell about your life):

> The ideal result of the process of belief relaxation and revision is a recalibration of the relevant beliefs so that they may better align or harmonize with other levels of the system and with bottom-up information—whether originating

> from within (e.g., via lower-level intrinsic systems and related interoception) or, at lower doses, outside the individual (i.e., via sensory input or extroception). Such functional harmony or realignment may look like a system better able to guide thought and behavior in an open, unguarded way (Watts et al., 2017; Carhart-Harris et al., 2018b).

This makes psychedelics a potent tool for psychotherapy:

> Consistent with the model presented in this work, overweighted high-level priors can be all consuming, exerting excessive influence throughout the mind and brain's (deep) hierarchy. The negative cognitive bias in depression is a good example of this (Beck, 1972), as are fixed delusions in psychosis (Sterzer et al., 2018). In this paper, we propose that psychedelics can be therapeutically effective, precisely because they target the high levels of the brain's functional hierarchy, primarily affecting the precision weighting of high-level priors or beliefs. More specifically, we propose that psychedelics dose-dependently relax the precision weighting of high-level priors (instantiated by high-level cortex), and in so doing, open them up to an upsurge of previously suppressed bottom-up signaling (e.g., stemming from limbic circuitry). We further propose that this sensitization of high-level priors means that more information can impress on them, potentially inspiring shifts in perspective, felt as insight. One might ask whether relaxation followed by revision of high-level priors or beliefs via psychedelic therapy is easy to see with functional (and anatomic) brain imaging. We presume that it must be detectable, if the right questions are asked in the right way.

Am I imagining this, or are Friston + Carhart-Harris and *Unlocking The Emotional Brain* getting at the same thing? Both start with a piece of a predictive model (= high-level prior) telling you something that doesn't fit the current situation. Both also assume you have enough evidence to convince a rational person that the high-level prior is wrong, or doesn't apply. But you don't automatically smash the prior and the evidence together and perform an update. In *UtEB*'s model, the update doesn't happen until you forge conscious links to both pieces of information and try to hold them in consciousness at the same time. In F+CH's model, the update doesn't happen until you take psychedelics which make the high-level prior lose some of its convincingness. *UtEB* is trying to laboriously build roads through mountains; F+CH are trying to cast a magic spell that makes the mountains temporarily vanish. Either way, you get communication between areas that couldn't communicate before.

III.

Why would mental mountains exist? If we keep trying to get rid of them, through therapy or psychedelics, or whatever, then why not just avoid them in the first place?

Maybe generalization is just hard (thanks to MC for this idea). Suppose Goofus is mean to you. You learn Goofus is mean; if this is your first social experience, maybe you also learn that the world is mean and people have it out for you. Then one day you meet Gallant, who is nice to you. Hopefully the system generalizes to "Gallant is nice, Goofus is still mean, people in general can go either way".

But suppose one time Gallant is just having a terrible day, and curses at you, and that time he happens to be wearing a red shirt. You don't want to overfit and conclude "Gallant wearing a red shirt is mean, Gallant wearing a blue shirt is nice". You want to conclude "Gallant is generally nice, but sometimes slips and is mean".

But any algorithm that gets too good at resisting the temptation to separate out red-shirt-Gallant and blue-shirt-Gallant risks falling into the opposite failure mode where it doesn't separate out Gallant and Goofus. It would just average them out, and conclude that people (including both Goofus and Gallant) are medium-niceness.

And suppose Gallant has brown eyes, and Goofus green eyes. You don't want your algorithm to overgeneralize to "all brown-eyed people are nice, and all green-eyed people are mean". But suppose the Huns attack you. You *do* want to generalize to "All Huns are dangerous, even though I can keep treating non-Huns as generally safe". And you want to do this as quickly as possible, definitely before you meet any more Huns. And the quicker you are to generalize about Huns, the more likely you are to attribute false significance to Gallant's eye color.

The end result is a predictive model which is a giant mess, made up of constant "This space here generalizes from this example, except this subregion, which generalizes from this other example, except over here, where it doesn't, and definitely don't *ever* try to apply any of those examples over here". Somehow this all works shockingly well. For example, I spent a few years in Japan, and developed a good model for how to behave in Japanese culture. When I came back to the United States, I effortlessly dropped all of that and went back to having America-appropriate predictions and reflexive actions (except for an embarrassing habit of bowing whenever someone hands me an object, which I still haven't totally eradicated).

In this model, mental mountains are just the context-dependence that tells me not to use my Japanese predictive model in America, and which prevents evidence that makes me update my Japanese model (like "I notice subways are always on time") from contaminating my American model as well. Or which prevent things I learn about Gallant (like "always trust him") from also contaminating my model of Goofus.

There's actually a real-world equivalent of the "red-shirt-Gallant is bad, blue-shirt-Gallant is good" failure mode. It's called "splitting"[9] and you can find it in any psychology textbook. Wikipedia defines it as "the failure in a person's thinking to bring together the dichotomy of both positive and negative qualities of the self and others into a cohesive, realistic whole".

In the classic example, a patient is in a mental hospital. He likes his doctor. He praises the doctor to all the other patients, says he's going to nominate her for an award when he gets out.

Then the doctor offends the patient in some way – maybe refuses one of his requests. All of a sudden, the doctor is abusive, worse than Hitler, worse than Mengele. When he gets out he will report her to the authorities and sue her for everything she owns.

Then the doctor does something right, and it's back to praise and love again.

The patient has failed to integrate his judgments about the doctor into a coherent whole, "doctor who sometimes does good things but other times does bad things". It's as if there's two predictive models, one of Good Doctor and one of Bad Doctor, and even though both of them refer to the same real-world person, the patient can only use one at a time.

Splitting is most common in borderline personality disorder. The DSM criteria for borderline includes splitting (there defined as "a pattern of unstable and intense interpersonal relationships characterized by alternating between extremes of idealization and devaluation"). They also include things like "markedly and persistently unstable self-image or sense of self", and "affective instability due to a marked reactivity of mood", which seem relevant here too.

Some therapists view borderline as a disorder of integration. Nobody is great at having all their different schemas talk to each other, but borderlines are atrocious at it. Their mountains are so high that even different thoughts about the same doctor can't necessarily talk to each other and coordinate on a coherent position. The capital only has enough messengers to talk to one valley at a time. If tribesmen from the Anger Valley are advising the capital today, the patient becomes truly angry, a kind of anger that utterly refuses to listen to any counterevidence, an anger pure beyond your imagination. If they are happy, they are *purely* happy, and so on.

About 70% of people[10] diagnosed with dissociative identity disorder (previously known as multiple personality disorder) have borderline personality disorder. The numbers are so high that some researchers[11] are not even convinced that these are two different conditions; maybe DID is just one manifestation of borderline, or especially severe borderline. Considering borderline as a failure of integration, this makes sense; DID is total failure of integration. People in the furthest mountain valleys, frustrated by inability to communicate meaningfully with the capital, secede and set up their own alternative provincial government, pulling nearby valleys into their new coalition. I don't want to overemphasize this; most popular perceptions of DID are overblown, and at least some cases seem to be at least partly iatrogenic. But if you are bad enough at integrating yourself, it seems to be the sort of thing that can happen.

In his review, Kaj relates this to Internal Family Systems, a weird form of therapy where you imagine your feelings as people/entities and have discussions with them. I've always been skeptical of this, because feelings are not, in fact, people/entities, and it's unclear why you should expect them to answer you when you ask them questions. And in my attempts to self-test the therapy, indeed nobody responded to my questions and I was left feeling kind of silly. But Kaj says:

> As many readers know, I have been writing a sequence of posts on multi-agent models of mind. In Building up to an Internal Family Systems model, I suggested that the human mind might contain something like subagents which try to ensure that past catastrophes do not repeat. In subagents, coherence, and akrasia in humans, I suggested that behaviors such as procrastination, indecision, and seemingly inconsistent behavior result from different subagents having disagreements over what to do.
>
> As I already mentioned, my post on integrating disagreeing subagents took the model in the direction of interpreting disagreeing subagents as conflicting beliefs or models within a person's brain. Subagents, trauma and rationality further suggested that the appearance of drastically different personalities within a single person might result from unintegrated memory networks, which resist integration due to various traumatic experiences.
>
> This post has discussed UtEB's model of conflicting emotional schemas in a way which further equates "subagents" with

> beliefs – in this case, the various schemas seem closely related to what e.g. Internal Family Systems calls "parts". In many situations, it is probably fair to say that this is what subagents are.

This is a model I can get behind. My guess is that in different people, the degree to which mental mountains form a barrier will cause the disconnectedness of valleys to manifest as anything from "multiple personalities", to IFS-findable "subagents", to *UtEB*-style psychiatric symptoms, to "ordinary" beliefs that don't cause overt problems but might not be very consistent with each other.

IV.

This last category forms the crucial problem of rationality.

One can imagine an alien species whose ability to find truth was a simple function of their education and IQ. Everyone who knows the right facts about the economy and is smart enough to put them together will agree on economic policy.

But we don't work that way. Smart, well-educated people believe all kinds of things, even when they should know better. We call these people biased, a catch-all term meaning something that prevents them from having true beliefs they ought to be able to figure out. I believe most people who don't believe in anthropogenic climate change are probably biased. Many of them are very smart. Many of them have read a lot on the subject (empirically, reading more about climate change will usually just make everyone more convinced[12] of their current position, whatever it is). Many of them have enough evidence that they should know better. But they don't.

(again, this is my opinion, sorry to those of you I'm offending. I'm sure you think the same of me. Please bear with me for the space of this example).

Compare this to Richard, the example patient mentioned above. Richard had enough evidence to realize that companies don't hate everyone who speaks up at meetings. But he still felt, on a deep level, like speaking up at meetings would get him in trouble. The evidence failed to connect to the emotional schema, the part of him that made the real decisions. Is this the same problem as the global warming case? Where there's evidence, but it doesn't connect to people's real feelings?

(maybe not: Richard might be able to say "I know people won't hate me for speaking, but for some reason I can't make myself speak", whereas I've never heard someone say "I know climate change is real, but for some reason I can't make myself vote to prevent it." I'm not sure how seriously to take this discrepancy).

In Crisis of Faith,[13] Eliezer Yudkowsky writes:

> Many in this world retain beliefs whose flaws a ten-year-old could point out, if that ten-year-old were hearing the beliefs for the first time. These are not subtle errors we're talking about. They would be child's play for an unattached mind to relinquish, if the skepticism of a ten-year-old were applied without evasion…we change our minds less often than we think.
>
> This should scare you down to the marrow of your bones. It means you can be a world-class scientist and conversant with Bayesian mathematics and still fail to reject a belief whose absurdity a fresh-eyed ten-year-old could see. It shows the invincible defensive position which a belief can create for itself, if it has long festered in your mind.

What does it take to defeat an error that has built itself a fortress?

He goes on to describe how hard this is, to discuss the "convulsive, wrenching effort to be rational" that he thinks this requires, the "all-out [war] against yourself". Some of the techniques he mentions explicitly come from psychotherapy,[14] others[15] seem to share a convergent evolution with it.

The authors of *UtEB* stress that all forms of therapy involve their process of reconsolidating emotional memories one way or another, whether they know it or not. Eliezer's work on crisis of faith feels like an *ad hoc* form of epistemic therapy, one with a similar goal.

Here, too, there is a suggestive psychedelic connection. I can't count how many stories I've heard along the lines of "I was in a bad relationship, I kept telling myself that it was okay and making excuses, and then I took LSD and realized that it obviously wasn't, and got out". Certainly many people change religions and politics[15] after a psychedelic experience, though it's hard to tell exactly what part of the psychedelic experience does this, and enough people end up believing various forms of woo that I hesitate to say it's all about getting more rational beliefs. But just going off anecdote, this sometimes works.

Rationalists wasted years worrying about various named biases, like the conjunction fallacy or the planning fallacy. But most of the problems we really care about aren't any of those. They're more like whatever makes the global warming skeptic fail to connect with all the evidence for global warming. If the model in *Unlocking The Emotional Brain* is accurate, it offers a starting point for understanding this kind of bias, and maybe for figuring out ways to counteract it.

(1) *Kaj Sotala. Book summary: Unlocking the Emotional Brain. 2019.*

(2) *Bruce Ecker, Robin Ticic, Laurel Hulley, Robert A. Neimeyer. 2012. Unlocking the Emotional Brain: Eliminating Symptoms at Their Roots Using Memory Reconsolidation.*

(3) *Nader, K., Schafe, G. E., & Le Doux, J. E. Fear memories require protein synthesis in the amygdala for reconsolidation after retrieval. 2000.*

(4) *Kellner, Charles H. MD; Romanella, Sara M. MS. ECT as a Novel Treatment for PTSD. Jun 2019.*

(5) *Scott Alexander. Book Review: All Therapy Books. 20th Nov 2019.*

(6) *R. L. Carhart-Harris, K. J. Friston. REBUS and the Anarchic Brain: Toward a Unified Model of the Brain Action of Psychedelics. Jul 2019.*

(7) *Scott Alexander. SSC Journal Club: Relaxed Beliefs Under Psychidelics And The Anarchic Brain. 10th Sept 2019.*

(8) *Wikipedia. Splitting (psychology).*

(9) *Paulette M. Gillig. Dissociative Identity Disorder. Mar 2009.*

(10) *J Lauer, D W Black, P Keen. Multiple personality disorder and borderline personality disorder. Distinct entities or variations on a common theme? Jun 1993.*

(11) *Maggie Koerth. Media Bubbles Aren't The Biggest Reason We're Partisans. 8th Oct 2019.*

(12) *Eliezer Yudkowsky. Crisis of Faith. 10th Oct 2008.*

(13) *Eliezer Yudkowsky. You Can Face Reality. 8th Aug 2007.*

(14) *Eliezer Yudkowsky. Original Seeing. 14th Oct 2007.*

(15) *Madison Margolin. Do Psychedelic Trips Change Your Political Views? 25th May 2017.*

Steelmanning Divination

The standard forms of divination that I've seen in contemporary Western culture— astrology, fortune cookies, lotteries, that sort of thing— seem pretty worthless to me. They're like trying to extract information from a random number generator, which is a generally hopeless phenomenon because of conservation of expected evidence.[1] Thus I had mostly written off divination; although I've come across some arguments[2] that divination served as a way to implement mixed strategies in competitive games. (Hunters would decide where to hunt by burning bones, which generated an approximately random map of their location, preventing their targets from learning where the humans liked to hunt and avoiding that location). But then I came across this striking passage, and sat up straight:

> One performs the rain sacrifice and it rains. Why? I say: there is no special reason why. It is the same as when one does not perform the rain sacrifice and it rains anyway. When the sun and moon suffer eclipse, one tries to save them. When Heaven sends drought, one performs the rain sacrifice. One performs divination and only then decides on important affairs. But this is not to be regarded as bringing one what one seeks, but rather is done to give things proper form. Thus, the gentleman regards this as proper form, but the common people regard it as connecting with spirits. If one regards it as proper form, one will have good fortune. If one regards it as connecting with spirits, one will have misfortune.

This is from Eric L. Hutton's translation of a collection of essays called Xunzi[3] (presumably written by Xunzi,[4] an ancient Chinese philosopher who was Confucian with heavy Legalist influences). The book was overall remarkable in how much of Xunzi's brilliance shone through, which is something I very rarely think about authors. (Talking to another rationalist who was more familiar with Chinese philosophy than I was, he also had this impression that Xunzi simply had a lot more mental horsepower than many other core figures). By the end of it, I was asking myself, "if they had *this much* of rationality figured out back then, why didn't they conquer the world"? Then I looked into the history a bit more and figured out that two[5] of[6] Xunzi's students were core figures in Qin Shi Huang's[7] unification of China to become the First Emperor.

So this paragraph stuck with me. When Xunzi talks about the way that earlier kings did things, I registered it as an applause light and moved on. When he talked about how an important role of government was to prevent innovation in music, I registered it as covering a very different thing than what I think of when I think about 'music' and moved on. But when he specifically called out the reason why I (and most educated people I know) don't pay much attention to astrology or other sorts of divination or magic, said "yeah, those would be dumb reasons to do this", and then said *"but there's still a reason"*, I was curious. What's the proper form that he's talking about? (Sadly, this was left as an exercise for the reader; the surrounding paragraphs are only vaguely related).

In his introduction, Hutton summarizes the relevant portion of Xunzi's philosophy:

> In this process of becoming good, ritual plays an especially important role in Xunzi's view. As he conceives them, the rituals constitute a set of standards for proper behavior that were created by the past sages and should govern virtually every aspect of a person's life. These rituals are not inviolable rules: Xunzi allows that people with developed moral judgment may need to depart from the strict dictates of ritual on some occasions, but he thinks those just beginning the process of moral learning need to submit completely to the requirements of ritual. Of the many important roles played by the rituals in making people good on Xunzi's view, three particularly deserve mention here. First the rituals serve to *display* certain attitudes and emotions. The ritually prescribed actions in the case of mourning, for instance, exhibit grief over the loss of a loved one, whether or not the ritual practitioner actually feels sadness. Second, even if the ritual practitioner does not actually feel the particular attitude or emotion embodied in the ritual, Xunzi believes that repeated performance of the ritual can, when done properly, serve to *cultivate* those attitudes and emotions in the person. To use a modern example, toddlers who do not know to be grateful when given a gift may be taught to say "thank you" and may do so without any understanding of its meaning or a feeling or gratitude. With repetition, time, and

a more mature understanding of the meaning of the phrase, many of these children grow into adults who not only feel gratitude upon receiving gifts but also say "thank you" as a conscious expression of that feeling. Similarly, on Xunzi's view, rituals serve to inculcate attitudes and feelings, such as caring and respect, that are characteristic of virtue, and then serve to express a person's virtue once it is fully developed. A third important function of the rituals is to allot different responsibilities, privileges, and goods to different individuals, and thereby help to prevent conflict over those things among people.

So what is cultivated by performing divination?

• • •

The first step is figuring out what sort of divination we're discussing. Xunzi probably had in mind the I Ching, a book with 64 sections, each corresponding to a situation or perspective, and advice appropriate for that situation. In the simplest version, one generates six random bits and then consults the appropriate chapter. I actually tried this for about a month, and then have done it off and on since then. I noticed several things about it that seemed useful:

- Entries in the I Ching typically focused on perspectives or principles instead of situations, consequences, or actions. Today's Taurus horoscope says "your self-esteem might be challenged by a fast talker or unpleasant situation" and counsels me "don't accept things as they appear on first glance", whereas the I Ching reading I just randomly selected talks about how following proper principles leads to increased power and how the increased power tempts us to abandon the principles that generated that power. This makes it much easier to scan one's life and see where the perspectives shed new light on a situation, or where principles had been ignored. (One of the early successes of my I Ching practice was a chapter that suggested reaching out to a trusted guide for advice, and I realized I should talk to a mentor at work about a developing situation, which I wouldn't have done otherwise).

- Given that, daily divination almost filled the same role as daily retrospectives or planning sessions; I was frequently thinking about all the different parts of my life on a regular interval, using a variety of random access to filter things down.
- "One performs divination and only then decides on important affairs". Often one is faced with a challenge that is "above one's pay grade", and having a prescribed ritual for what sort of cognition needs to be done encourages reflection and popping out of the obvious frame. Simply thinking about a situation in the way one naturally would doesn't correct for biases, while attempting to make sense of a situation from a randomly generated frame does help expand one's conception of it.
- Since the divination result was a particular perspective (rather than an object-level claim about events or the correct action), it was easy to see when the perspective had been 'fully considered' and the reflection was done. If I'm trying to make a binary choice and I flip a coin to resolve it, basically all I'm doing is checking whether or not my gut is secretly hoping the coin lands a particular way, and in cases of genuine uncertainty I will end up just as uncertain after consulting the coin flip. But when I have a situation and I resolve the questions of "what does patience have to say about X"? and "what does humility have to say about X"? then I can have the sense of having actually made progress.
- It encouraged experimentation by partially decoupling one's mood and one's decisions. The first few days, I was instructed to consider diligence and so I worked more than I would have wanted to (and discovered that this was generally fine); on the fourth day, I joked, "when is it going to tell me to goof off and play video games"?, and then got a reading that said "effort is hopeless today; that happens, cope with it". So despite feeling

like I could have been productive, I took the day off, and later had the sense that I had confirmed my initial sense that I didn't need to take the day off, providing data to calibrate on that I wouldn't have gotten except through random variation.

Essentially, it looked to me like the steelman of divination is something like Oblique Strategies,[8] where challenging situations (either 'daily life' or a specific important decision) are responded to by random access to a library of perspectives or approaches, and the particular claim made by a source is what distribution is most useful. There was previously an attempt on LW[9] to learn what advice was useful, but I think on the wrong level of abstraction (the 'do X' variety, instead of the 'think about X' variety).

This approach has also served me well with other forms of divination I've since tried; a Tarot deck works by focusing your attention on a situation, and then randomly generating a frame (from an implied distribution of relevant symbols), giving one access to parts of the space that they wouldn't have considered otherwise. This also trains the habit of 'understanding alien frames'; if I am considering a conflict with another person and then have to figure out what it means that "I'm the vizier of water, the relationship is the three of earth, and the other person is strength" (where, of course, each of those is in fact an image rich in detail rather than a simple concept), this trains the habit of adopting other perspectives / figuring out how things make sense from the other point of view.

(1) *EliezerYudkowsky. Conservation of Expected Evidence. 13 Aug 2007.*

(2) *Omar Khayyam Moore. Divination-A New Perspective.*

(3) *Wikipedia. Xunzi (book).*

(4) *Wikipedia. Xun Kuang.*

(5) *Wikipedia. Li Si.*

(6) *Wikipedia. Han Fei.*

(7) *Wikipedia. Qin Shi Huang.*

(8) *Wikipedia. Oblique Strategies.*

(9) *Adelene Dawner. Rationalist horoscopes: A low-hanging utility generator. 22 May 2011.*

The Engines of Cognition:
Essays by the LessWrong Community

Published in the United States of America
by LessWrong Press.
Berkeley, California
LessWrong.com

Printed by Hemlock in Canada
Designed by Asimov Collective in Brooklyn, New York
ISBN: 978-1-7361285-1-0
First edition

To be in a future year's set of books,
publish essays on LessWrong.com.